Python
全栈开发 Web 编程

夏正东 ◎ 编著

清华大学出版社
北京

内 容 简 介

本书重点讲解Web编程的相关知识点,并搭配了150多个示例代码和两个综合项目,可以帮助读者快速、深入地理解和应用相关技术。

本书共6章。第1章Web编程简介,简要介绍使用Python进行Web编程的若干知识点;第2章Web编程的常用概念,主要介绍Web开发过程涉及的相关概念;第3章Flask,包括Flask简介、安装Flask、第1个Flask项目、Flask实例对象、路由、模板、类视图、蓝图、SQLAlchemy、Flask-SQLAlchemy、Alembic、Flask-Script、Flask-Migrate、表单验证、Cookie和Session、上下文、钩子函数和信号等内容;第4章Flask项目实战:网上图书商城,包括程序概述、创建数据库、程序目录结构和程序编写等知识点;第5章Django,包括Django简介、安装Django、第1个Django项目、路由、模板、类视图、数据库、表单验证、Cookie和Session、上下文处理器、中间件和CSRF防御等;第6章Django项目实战:网上图书商城,包括程序概述、数据库设计和编写程序等内容。

本书可以作为广大计算机软件技术人员的参考用书,也可以作为高等院校计算机科学与技术、自动化、软件工程、网络工程、人工智能和信息管理与信息系统等专业的教学参考用书。

版权所有,侵权必究。举报:010-62782989,beiqinquan@tup.tsinghua.edu.cn。

图书在版编目(CIP)数据

Python全栈开发:Web编程 / 夏正东编著. -- 北京:清华大学出版社,2025.5.
(清华开发者书库). -- ISBN 978-7-302-69344-4

Ⅰ.TP312.8

中国国家版本馆CIP数据核字第2025RC0117号

责任编辑:赵佳霓
封面设计:刘　键
责任校对:时翠兰
责任印制:宋　林

出版发行:清华大学出版社
网　　址:https://www.tup.com.cn,https://www.wqxuetang.com
地　　址:北京清华大学学研大厦A座　　邮　编:100084
社 总 机:010-83470000　　邮　购:010-62786544
投稿与读者服务:010-62776969,c-service@tup.tsinghua.edu.cn
质量反馈:010-62772015,zhiliang@tup.tsinghua.edu.cn
课件下载:https://www.tup.com.cn,010-83470236

印 装 者:三河市君旺印务有限公司
经　　销:全国新华书店
开　　本:185mm×260mm　　印　张:19　　字　数:464千字
版　　次:2025年7月第1版　　　　　　　印　次:2025年7月第1次印刷
印　　数:1~1500
定　　价:79.00元

产品编号:093016-01

序
FOREWORD

Python 自公开发布以来已有 30 多年的历史，近几年更成为炙手可热的编程语言。在多数知名技术交流网站的排名中，能稳定地排在前 3 名，说明了 Python 的巨大市场需求和良好的发展前景，也使更多人希望学习和掌握 Python 编程技术，以便提升自身的竞争力，乃至获得更好的求职机会。

Python 语言的流行得益于自身的特点和能力。首先，作为一种通用语言，Python 具有简单、易学、免费、开源、可移植、可扩展、可嵌入和面向对象等诸多优点，能帮开发者轻松地完成编程工作；其次，Python 被广泛地应用于 GUI 设计、游戏编程、Web 开发、运维自动化、科学计算、数据可视化、数据挖掘及人工智能等多行业和领域。有专业调查显示，Python 已成为越来越多开发者的语言选择。目前，国内外很多大企业和机构（如豆瓣、搜狐、金山、腾讯、网易、百度、阿里巴巴、淘宝、谷歌、YouTube 和 Facebook 等）在应用 Python 完成各种各样的任务。

时至今日，Python 几乎可以应用于任何领域和场合。

从近几年相关领域招聘岗位的需求来看，Python 工程师岗位的需求量巨大，并且还在呈现不断上升的趋势。根据知名招聘网站的数据显示，目前全国 Python 岗位的需求量接近 10 万个，平均薪资水平约在 13 000 元。可见，用"炙手可热"描述 Python 工程师并不为过。

本书是市面上难得一见的同时汇集了 Flask 和 Django 全开发流程的 Python 高阶图书。除了涵盖知识点广泛的特点之外，其内容编排也非常新颖，各章节之间既有独立性，又能递进支撑，可以有效地缩短学习的时间。此外，本书搭配了多个示例代码和综合项目，使原本比较难以理解和学习的 Web 编程变得容易接受，极大地提升了读者的学习乐趣和信心。

本书的另一个值得推荐的理由来自作者的工程素养。与一般的高阶技术书籍不同，本书在讲述语法和编程知识的同时，更认真、更细致地介绍了与工程相关的规范，并且这种规范贯穿了示例代码的始终。对于实际的软件开发工作来讲，它们既是必须掌握的知识，又是在实际编程实践中应具备的良好素养。

衷心希望广大读者通过本书快速掌握 Web 编程的相关技术，体会到运用 Python 解决工作中的实际问题所带来的乐趣和成就感。同时，也希望作者能够再接再厉，为广大读者奉献更多的优质书籍。

牛连强
2025 年 2 月于沈阳工业大学

前 言
PREFACE

随着互联网时代的到来，众多编程语言走进了大众的视野。在当前大数据、人工智能方兴未艾之时，相关工作岗位所需要的技术人才更是一度出现供不应求的现象，而 Python 正是应用于上述技术领域的最佳编程语言。

Python 横跨多个互联网核心技术领域，并且以其简单高效的特点，被广泛地应用于各种应用场景，包括 GUI 开发、游戏开发、Web 开发、运维自动化、科学计算、数据可视化、数据挖掘及人工智能等。

此外，随着国家对人工智能等技术领域的重视和布局，更凸显出 Python 的重要地位。从 2018 年起，浙江省信息技术教材已启用 Python，放弃 Visual Basic，这一改动也意味着 Python 将成为浙江高考内容之一。更有前瞻性的是，山东省最新出版的小学信息技术教材，在六年级信息技术课本中也加入了 Python 的相关内容——终于小学生也开始学习 Python 了。

而本书正是在这样的背景之下笔者的倾情之作。本书是 Python 全栈开发系列的第 4 册，全书共分为 6 章，将重点讲解主流的 Web 开发框架，即 Flask 和 Django，并搭配 150 多个示例代码和两个综合项目，理论知识与实战开发并重，可以帮助读者快速、深入地理解和应用 Web 编程的相关技术。扫描目录上方的二维码可下载本书源码。

著名华人经济学家张五常曾经说过，"即使世界上 99% 的经济学论文没有发表，世界依然会发展成现在这样子"，而互联网时代的发展同样具有其必然性，所以要想成功，我们就必须顺势而为，真正地站稳在时代的风口之上。

衷心致谢

首先，感谢每位读者，感谢您在茫茫书海中选择了本书，笔者衷心地祝愿各位读者能够借助本书学有所成，并最终顺利地完成自己的学习目标、学业考试和职业选择。

其次，感谢笔者的导师、同事、学生和朋友，感谢他们不断地鼓励和帮助笔者，非常荣幸能够和这些聪明、勤奋、努力、踏实的人一起学习、工作和交流。

最后，感谢笔者的父母，是他们给予了我所需要的一切，没有他们无私的爱，就没有笔者今天的事业，更不能达成我的人生目标。

此外，本书在编写和出版过程中得到了来自沈阳工业大学牛连强教授、大连东软信息学院张明宝副教授、大连华天软件有限公司陈秋男先生、51CTO 学堂曹亚莉女士、印孚瑟斯技术(中国)有限公司崔巍先生和清华大学出版社赵佳霓编辑的大力支持和帮助，在此衷心地表示感谢。

在本书的编写过程中，笔者虽然始终本着科学、严谨的态度，力求精益求精，但书中难免存在疏漏之处，恳请广大读者批评指正。

夏正东

2025 年 5 月 1 日于辽宁大连

目 录
CONTENTS

本书源码

第 1 章 Python Web 编程简介 ·· 1

第 2 章 Python Web 编程的常用概念 ··· 3

第 3 章 Flask ··· 9

  3.1 Flask 简介 ··· 9
  3.2 安装 Flask ··· 9
  3.3 第 1 个 Flask 项目 ·· 9
  3.4 Flask 实例对象 ··· 11
  3.5 路由 ·· 16
      3.5.1 路由的声明 ··· 16
      3.5.2 唯一规则 ··· 17
      3.5.3 路由分类 ··· 17
      3.5.4 动态构建请求 URL ·· 19
      3.5.5 HTTP 请求 ·· 20
      3.5.6 路由转换器 ··· 22
      3.5.7 重定向 ··· 25
      3.5.8 视图函数的返回值 ·· 26
      3.5.9 自定义视图函数装饰器 ·· 28
  3.6 模板 ·· 29
      3.6.1 渲染模板 ··· 29
      3.6.2 模板位置 ··· 30
      3.6.3 模板变量 ··· 31
      3.6.4 模板中动态构建请求 URL ··· 32
      3.6.5 模板中的过滤器 ··· 33
      3.6.6 模板中的控制结构 ·· 38
      3.6.7 模板注释 ··· 43

    3.6.8 宏 ·· 44
    3.6.9 include 标签 ··· 46
    3.6.10 set 语句和 with 语句 ··· 48
    3.6.11 加载静态文件 ·· 49
    3.6.12 模板继承 ·· 51
3.7 类视图 ··· 53
3.8 蓝图 ··· 58
    3.8.1 应用蓝图 ·· 58
    3.8.2 蓝图中加载模板 ·· 60
    3.8.3 蓝图中加载静态文件 ·· 63
    3.8.4 在蓝图中动态构建请求 URL ·· 64
    3.8.5 在蓝图中实现子域名 ·· 67
3.9 SQLAlchemy ··· 68
    3.9.1 安装 SQLAlchemy ·· 68
    3.9.2 创建数据库引擎 ·· 68
    3.9.3 创建数据库 ·· 69
    3.9.4 创建数据表 ·· 69
    3.9.5 CRUD 操作 ·· 71
    3.9.6 外键 ·· 80
    3.9.7 多表间关系 ·· 81
    3.9.8 高级查询 ·· 87
3.10 Flask-SQLAlchemy ·· 97
    3.10.1 安装 Flask-SQLAlchemy ·· 97
    3.10.2 配置 Flask-SQLAlchemy ·· 97
    3.10.3 连接数据库 ·· 98
    3.10.4 获取数据库对象 ·· 98
    3.10.5 创建数据表 ·· 99
    3.10.6 CRUD 操作 ·· 100
    3.10.7 多表间关系 ·· 104
3.11 Alembic ··· 108
    3.11.1 安装 Alembic ··· 108
    3.11.2 Alembic 操作 ··· 108
    3.11.3 在 Flask-SQLAlchemy 中操作 Alembic ································ 112
3.12 Flask-Script ··· 113
    3.12.1 安装 Flask-Script ·· 114
    3.12.2 创建自定义命令 ·· 114
3.13 Flask-Migrate ·· 117
    3.13.1 安装 Flask-Migrate ··· 117
    3.13.2 Flask-Migrate 操作 ··· 117

- 3.14 表单验证 ·· 119
  - 3.14.1 安装 WTForms 和 Flask-WTF ··· 119
  - 3.14.2 HTML 表单验证 ··· 119
  - 3.14.3 文件上传验证 ·· 123
- 3.15 Cookie 和 Session ·· 126
  - 3.15.1 设置、获取和删除 Cookie ·· 126
  - 3.15.2 设置、获取和删除 Session ··· 128
- 3.16 CSRF 防御 ··· 129
- 3.17 上下文 ··· 131
  - 3.17.1 应用上下文 ·· 131
  - 3.17.2 请求上下文 ·· 132
  - 3.17.3 应用上下文和请求上下文的区别 ··· 133
- 3.18 钩子函数 ··· 134
- 3.19 信号 ··· 138
  - 3.19.1 信号的安装 ·· 139
  - 3.19.2 自定义信号 ·· 139
  - 3.19.3 内置信号 ·· 140

第 4 章 Flask 项目实战：网上图书商城 ·· 144

- 4.1 程序概述 ··· 144
- 4.2 创建数据库 ··· 146
- 4.3 程序目录结构 ··· 149
- 4.4 程序编写 ··· 149

第 5 章 Django ·· 167

- 5.1 Django 简介 ·· 167
- 5.2 安装 Django ·· 167
- 5.3 第 1 个 Django 项目 ··· 167
- 5.4 路由 ··· 173
  - 5.4.1 视图函数 ··· 173
  - 5.4.2 URL 映射 ··· 174
  - 5.4.3 HttpRequest 对象 ··· 175
  - 5.4.4 QueryDict 对象 ··· 176
  - 5.4.5 HttpResponse 对象 ·· 176
  - 5.4.6 JsonResponse 对象 ·· 177
  - 5.4.7 重定向 ··· 177
  - 5.4.8 动态路由 ··· 177
  - 5.4.9 动态构建请求 URL ··· 178
  - 5.4.10 路由分发 ·· 179

	5.4.11 路由转换器	183
	5.4.12 限制请求方法	187
5.5	模板	189
	5.5.1 渲染模板	189
	5.5.2 模板位置	191
	5.5.3 模板变量	194
	5.5.4 模板中的控制结构	195
	5.5.5 模板注释	201
	5.5.6 常用标签	202
	5.5.7 模板中的过滤器	208
	5.5.8 模板继承	212
	5.5.9 加载静态文件	214
5.6	类视图	219
5.7	数据库	221
	5.7.1 定义数据模型	222
	5.7.2 Manager 类和 QuerySet 类	224
	5.7.3 查询条件	225
	5.7.4 常用字段	234
	5.7.5 Meta 类	234
	5.7.6 外键	234
	5.7.7 多表间关系	242
5.8	表单验证	250
	5.8.1 HTML 表单验证	250
	5.8.2 上传文件验证	254
	5.8.3 ModelForm 类	255
5.9	Cookie 和 Session	259
	5.9.1 设置、获取和删除 Cookie	259
	5.9.2 设置、获取和删除 Session	261
5.10	上下文处理器	264
5.11	中间件	267
5.12	CSRF 防御	269

第 6 章 Django 项目实战：网上图书商城 272

6.1	程序概述	272
6.2	数据库设计	273
6.3	编写程序	275

第 1 章 Python Web 编程简介

随着互联网技术的普及，整个信息产业得到了迅猛发展。在信息产业技术中，Web 应用开发是一项占有重要地位的应用技术。

当前，国内外的信息化建设已经进入了基于 Web 应用为核心的阶段，各种信息管理系统的工程架构模式逐步由传统的单机模式或 C/S 架构模式向 B/S 架构模式转变，而在主流的 Web 技术中，Python 的开发速度和灵活性使其可以很容易地快速建立和运行网站，并且由于 Python 语言能够满足快速迭代的需求，所以非常适合互联网公司的 Web 开发应用场景。

Python 用于 Web 应用开发已有十余年的历史，在这个过程中，涌现出了众多优秀的 Web 开发框架，许多知名网站也在使用 Python 语言进行 Web 应用开发，例如豆瓣、知乎、网易、YouTube 和 Yelp 等。这一方面足以说明 Python 用于 Web 应用开发的受欢迎程度，另一方面也说明 Python 语言用于 Web 应用开发已经经受住了大规模用户并发访问的考验。

Python 的 Web 编程涵盖多个方面，其不仅包含服务器端编程，还涉及数据库管理和前端设计等。

1. 服务器端编程

在服务器端编程领域，Python 拥有多种强大的 Web 开发框架。

Python Web 开发框架是用于简化 Web 编程开发流程的软件框架。它不是一项全新的技术，而是一些常用功能的集合，以达到降低 Web 编程开发成本的目的。

Python Web 开发框架的常用功能包括管理路由、支持数据库、支持 MVC 模式、支持 ORM、支持模板引擎，以及管理会话等。

常用的 Python Web 开发框架包括 Flask、Django、Tornado、Twisted 和 FastAPI 等。

（1）Flask 是一个轻量级的可定制框架，使用 Python 语言编写，较其他同类型框架更灵活、轻便、安全且容易上手。它可以很好地结合 MVC 模式进行开发，经过开发人员的合理分工合作，小型团队便可以在短时间内完成功能丰富的中小型网站。

Flask 还有很强的定制性，用户可以根据自己的需求来添加相应的功能，在保持核心功能简单的同时，实现功能的丰富与扩展。此外，其强大的插件库可以帮助用户实现个性化的网站定制，以开发出功能强大的网站。

（2）Django 是一个由 Python 编写的具有完整架站能力的开源 Web 开发框架。通过 Django 框架，只需很少的代码，开发人员就可以轻松地完成一个正式网站所需要的大部分内容,并进一步开发出全功能的Web服务。

Django遵循MVC设计模式,具有开发快捷、部署方便、可重用性高和维护成本低等优点。

(3) Tornado是一个使用Python编写的非阻塞式Web服务器软件和框架,由FriendFeed开发并由Facebook维护。

Tornado最初是为了处理FriendFeed的实时功能而开发的,并且由于其具有非阻塞的特性,所以能够处理大量的并发请求,特别适合需要高性能实时交互的Web应用。

Tornado的主要特点包括非阻塞式服务器、高效的长连接支持、协程支持、轻量级和高可扩展性,以及内置HTTP服务器。

Tornado的应用场景包括高并发实时服务和长连接应用等。

(4) Twisted是一个完整的事件驱动的网络编程框架,允许开发者构建高性能、可伸缩的网络应用和协议。

Twisted支持多种网络协议,例如HTTP、SMTP和IMAP等,并且提供了丰富的工具和组件来构建服务器、客户端和聊天应用等。

Twisted的核心特点是其异步事件驱动模型,其使用事件循环来处理网络事件,可以同时处理多个连接而不会阻塞主程序。这种模型使Twisted非常适合开发高性能的网络应用。

此外,Twisted的组件是模块化和可扩展的,开发者可以选择性地使用不同的模块和插件来满足项目的需求。

(5) FastAPI是一个现代、快速且高性能的Web开发框架,用于构建基于Python的API。

FastAPI基于Starlette和Pydantic构建,支持异步编程,具有高性能和低延迟的特点。

FastAPI使用标准的Python类型提示,使代码更加规范、可读,并通过类型检查工具,例如Mypy进行静态类型检查,以提高代码质量。

FastAPI的特性包括异步支持、类型提示、自动生成文档、数据验证与序列化、依赖注入系统、类型检查、中间件支持和标准化等。

2. 数据库管理

数据库管理在Web编程中至关重要。Python提供了SQLAlchemy和psycopg2等数据库链接库,使之与MySQL、PostgreSQL和SQLite等主流数据库的交互变得更加轻松与便捷。

3. 前端设计

Python在前端设计方面也表现出色。拥有Jinja2和DTL等优秀模板引擎,可协助开发者生成动态且响应式的HTML内容。

第 2 章 Python Web 编程的常用概念

在正式学习 Web 编程之前,读者需要了解一些在 Web 开发过程中涉及的重要概念,即 MVC 模式、ORM、WSGI 和虚拟环境。

1. MVC 模式

MVC 模式最早由 Trygve Reenskaug 在 1978 年提出,是施乐帕克研究中心(Xerox PARC)在 20 世纪 80 年代为程序语言 Smalltalk 发明的一种软件设计模式。

MVC 模式的主要目的包括三点,一是将数据的处理和数据的展示分开;二是使类的维护和实现更加简单;三是灵活地改变数据的存储和显示方式,使两者相互独立。

使用 MVC 模式最终是为了实现一种动态的程序设计,进而使后续对程序的修改和扩展简化,以及使程序某一部分的重复利用成为可能。

通常 MVC 模式将 Web 应用程序分为 3 个基本部分,即模型(Model)、视图(View)和控制器(Controller)。这 3 部分相互关联,有助于将数据的处理与数据的呈现分开。

1)模型

主要用于提供数据和业务逻辑,即定义针对数据的所有操作(例如创建、修改和删除等),并提供与数据使用相关的方法。

2)视图

主要用于数据的展示,即提供相关的方法,以便根据上下文和应用程序的需要构建 Web 界面。

3)控制器

主要用于从请求中接收数据,并将其发送到系统的其他部分。

综上所述,模型是应用程序的基石,由于提供客户端请求的数据,所以必须在多个操作中保持一致;视图用于将数据展示在接口上以供用户查看,所以可以独立开发,但不应包含复杂的逻辑,相反其需要足够灵活,以适应多种平台,并且应避免与数据库直接交互;控制器作为模型和视图之间的黏合剂,不应该进行数据库调用或参与数据的展示。

2. ORM

对象关系映射(Object Relational Mapping,ORM)是一种用于解决面向对象和关系数据库之间数据交互问题的技术。这样,在具体操作业务对象时,就不需要再去和复杂的 SQL 语句打交道,只需简单地操作对象的属性和方法,所以 ORM 本质上就是一种用面向对象的思维来操作关系数据库中数据的技术。

众所周知,关系数据库是企业级应用环境中永久存放数据的主流数据存储系统,而对象和关系数据是业务实体的两种表现形式,即业务实体在内存中表现为对象,在数据库中则表现为关系数据,但是在内存中,对象之间存在关联和继承关系,而在数据库中,关系数据则无法直接表达多对多关联和继承关系。

例如,创建一个 User 类,其属性 id、username 和 password 都可以用来记录该用户的信息,而当需要把该对象中的数据存储到数据库中时,按照传统思路,则需要手动编写 SQL 语句,并将对象的属性值提取到 SQL 语句中,然后调用相关方法执行 SQL 语句。此时,手动编写 SQL 语句的缺点非常明显,主要体现在两个方面:一是由于对象的属性名和数据表的字段名往往不一致,所以在编写 SQL 语句时需要非常细致,要逐一核对属性名和数据表的字段名,确保其不出错,并且彼此之间要一一对应;二是当 SQL 语句出错时,数据库的提示信息往往也不精准,这就给排除错误带来了较大的困难。

而 ORM 的出现,恰好解决了这些难题。ORM 可以使两者之间的交互变得自动化,即只需提前配置好对象和数据之间的映射关系,ORM 就可以自动地生成 SQL 语句,并将对象中的数据自动存储到数据库中,整个过程无须人工干预,在解放了程序员双手的同时,也让源代码中不再出现 SQL 语句。

需要说明的是,ORM 是一种双向数据交互技术,它不仅可以将对象中的数据存储到数据库中,也可以反过来将数据库中的数据提取到对象中。

但是,ORM 在提高开发效率的同时,也带来了几个缺点:一是 ORM 增加了学习成本;二是自动生成 SQL 语句会消耗计算资源,这势必会对程序性能造成一定的影响;三是对于复杂的数据库操作,ORM 通常难以处理,即使可以处理,自动生成的 SQL 语句在性能方面也不如手动编写的原生 SQL 语句;四是由于生成 SQL 语句的过程是自动进行的,不能人工干预,所以这使开发人员无法定制一些特殊的 SQL 语句。

3. WSGI

众所周知,在处理 HTTP 请求时,Web 服务器需要与 Web 应用程序进行通信,但是由于 Web 服务器和 Web 应用程序的种类繁多,如果不统一标准,就会存在 Web 服务器和 Web 应用程序无法匹配的情况,进而导致开发受到限制,而为了有效地解决该问题,WSGI 就应运而生了。

WSGI 是 Web Server Gateway Interface 的缩写,即 Web 服务器网关接口,是为 Python 语言定义的 Web 服务器和 Web 应用程序之间的标准接口,以提高 Web 应用程序在一系列 Web 服务器间的移植性。

综上所述,WSGI 存在的目的主要包括两个方面:一是使 Web 服务器知道如何调用 Web 应用程序,并且将客户端的 HTTP 请求传递给 Web 应用程序;二是使 Web 应用程序能够理解客户端的 HTTP 请求并执行对应操作,以及将执行结果返给 Web 服务器,并最终得到客户端的响应。

下面实现一个最简单的 Web 编程,以帮助读者更好地理解 WSGI。

(1) 创建一个符合 WSGI 标准的 HTTP 处理函数,用于实现 Web 应用程序,并响应 HTTP 请求,代码如下:

```
# 资源包\Code\chapter2\1\app.py
def application(environ, start_response):
```

```
start_response('200 OK', [('Content-Type', 'text/html')])
return [b'<h1>This is Web!</h1>']
```

其中,application()函数中的参数 environ 包含 HTTP 请求的所有信息,而参数 start_response 则是用于发起 HTTP 响应的函数。

通过上述代码,可以发现 application()函数本身并没有涉及任何解析 HTTP 的部分,也就是说,底层代码并不需要开发人员进行编写,开发人员只需负责在更高层次上考虑如何响应请求。

(2) 使用 Python 内置的符合 WSGI 规范的 Web 服务器调用该 Web 应用程序,代码如下:

```
# 资源包\Code\chapter2\1\server.py
from wsgiref.simple_server import make_server
# 导入 application 函数
from app import application
# 创建一个服务器,IP 地址为空,端口是 8000,处理函数是 application
httpd = make_server('', 8000, application)
print('Serving HTTP on port 8000...')
# 开始监听 HTTP 请求
httpd.serve_forever()
```

(3) 运行 server.py 文件,启动 Web 服务器,PyCharm 中的运行结果如图 2-1 所示。

(4) 当 Web 服务器启动成功后,打开浏览器并在网址栏输入 http://localhost:8000/,其显示内容如图 2-2 所示。

图 2-1　启动 Web 服务器　　　　图 2-2　浏览器中显示的内容

4. 虚拟环境

在实际开发过程中,经常会应用大量的第三方库或者模块,并且随着时间的推移,这些第三方库或模块的版本也会不断地迭代,此时就会遇到版本依赖的问题。

而虚拟环境包含了特定的 Python 解析器及一些软件包,可以使不同的应用程序使用不同的开发环境,从而解决了依赖冲突问题。此外,由于在虚拟环境中可以只安装与应用程序相关的包或者模块,这就给部署带来了极大的便利。

例如,在计算机中存在 A、B 两个项目,其使用同一个第三方库,并且运行一切顺利,但由于某种特殊原因,B 项目要应用这个第三方库的新版本特性,而此时如果贸然将第三方库升级至新版本,则 B 项目虽然可以顺利运行,但是对 A 项目的运行影响将是无法评估的,所以,此时就可以采用虚拟环境让 A 和 B 两个项目分别处于两个不同的 Python 开发环境中,使其互不影响。

1) 使用 virtualenv 搭建虚拟环境

（1）安装：由于 virtualenv 属于 Python 的第三方库，所以需要进行安装，打开命令提示符，输入命令 pip install virtualenv 即可。

（2）创建虚拟环境：在命令提示符中，通过命令"virtualenv 虚拟环境名"，即可完成虚拟环境的创建，如图 2-3 和图 2-4 所示。

图 2-3　创建虚拟环境 env1

图 2-4　虚拟环境 env1

此外，还可以通过"-p"指定虚拟环境所使用的 Python 版本，如图 2-5 所示。

图 2-5　创建指定 Python 版本的虚拟环境 env1_python37

（3）激活虚拟环境：在命令提示符中，进入所创建虚拟环境中的 Scripts 目录，并输入命令 activate，即可激活当前虚拟环境，如图 2-6 所示。

（4）退出虚拟环境：在当前虚拟环境中，输入命令 deactivate，即可退出当前虚拟环境，如图 2-7 所示。

图 2-6　激活虚拟环境 env1

图 2-7　退出虚拟环境 env1

2) 使用 virtualenvwrapper 搭建虚拟环境

（1）安装：由于 virtualenvwrapper 属于 Python 的第三方库，所以需要进行安装，打开命令提示符，输入命令 pip install virtualenvwrapper-win 即可。

（2）创建虚拟环境：在命令提示符中，通过命令"mkvirtualenv 虚拟环境名"，即可完成虚拟环境的创建，如图 2-8 所示。

此外，还可以通过"--python"指定虚拟环境所使用的 Python 版本，如图 2-9 所示。

这里需要注意，使用 virtualenvwrapper 搭建的虚拟环境，其默认路径为 C:\Users\Administrator\Envs，如图 2-10 所示。

图 2-8　创建虚拟环境 env2

图 2-9　创建指定 Python 版本的虚拟环境 env2_python37

图 2-10　虚拟环境的默认位置

如果需要更改虚拟环境的默认路径，则需要在计算机的环境变量中添加系统变量 WORKON_HOME，其值就是自定义的虚拟环境路径，如图 2-11 所示。

图 2-11　自定义虚拟环境路径

此时，重新打开命令提示符，并输入命令"mkvirtualenv 虚拟环境名"，即可在自定义的路径中创建虚拟环境，如图 2-12 所示。

图 2-12　创建虚拟环境 env2

（3）切换并激活虚拟环境：在命令提示符中，输入命令"workon 虚拟环境名"，即可切换并激活虚拟环境，如图 2-13 所示。

（4）退出虚拟环境：在当前虚拟环境中，输入命令 deactivate，即可退出当前虚拟环境，如图 2-14 所示。

（5）列出所有虚拟环境：在命令提示符中，输入命令 lsvirtualenv，即可列出所有虚拟环境，如图 2-15 所示。

图 2-13　切换并激活虚拟环境 env2　　　　图 2-14　退出虚拟环境

图 2-15　所有虚拟环境

（6）删除虚拟环境：在虚拟环境中，输入命令"rmvirtualenv 当前虚拟环境名"，即可删除虚拟环境，如图 2-16 所示。

图 2-16　删除虚拟环境 env2

第 3 章 Flask

3.1 Flask 简介

Flask 是一个使用 Python 编写的轻量级 Web 开发框架,其 WSGI 函数库采用 Werkzeug,而模板引擎则使用 Jinja2,较其他同类型框架更灵活、轻便、安全且容易上手。

此外,Flask 可以很好地结合 MVC 模式进行开发,使开发人员在短时间内就可以完成功能丰富的中小型网站。

Flask 还具有很强的定制性,用户可以根据自己的需求来添加相应的功能,即在保持核心功能简单的同时,实现功能的丰富与扩展,并且其强大的插件库可以让用户实现个性化的网站定制,以确保开发出功能强大的网站。

3.2 安装 Flask

由于 Flask 属于 Python 的第三方库,所以需要进行安装,打开命令提示符,输入命令"pip install flask"即可。

3.3 第 1 个 Flask 项目

通过 PyCharm(Professional)可以快速地创建一个 Flask 项目,具体步骤如下:

(1) 打开 PyCharm(Professional),选择 File→New Project,如图 3-1 所示。

(2) 首先在当前界面的左侧栏中选择 Flask,然后在右侧栏中单击 Previously configured interpreter,并在其中选择虚拟环境 flask_env,最后单击 Create 按钮,即可创建一个名为 flaskProject 的项目,如图 3-2 所示。

使用 PyCharm(Professional)创建项目 flaskProject,其结构如图 3-3 所示。

其中,文件 app.py 是 Flask 项目的主程序,文件夹

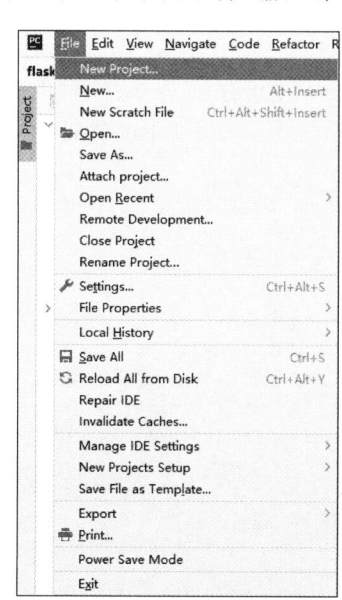

图 3-1 创建 Flask 项目的第 1 步

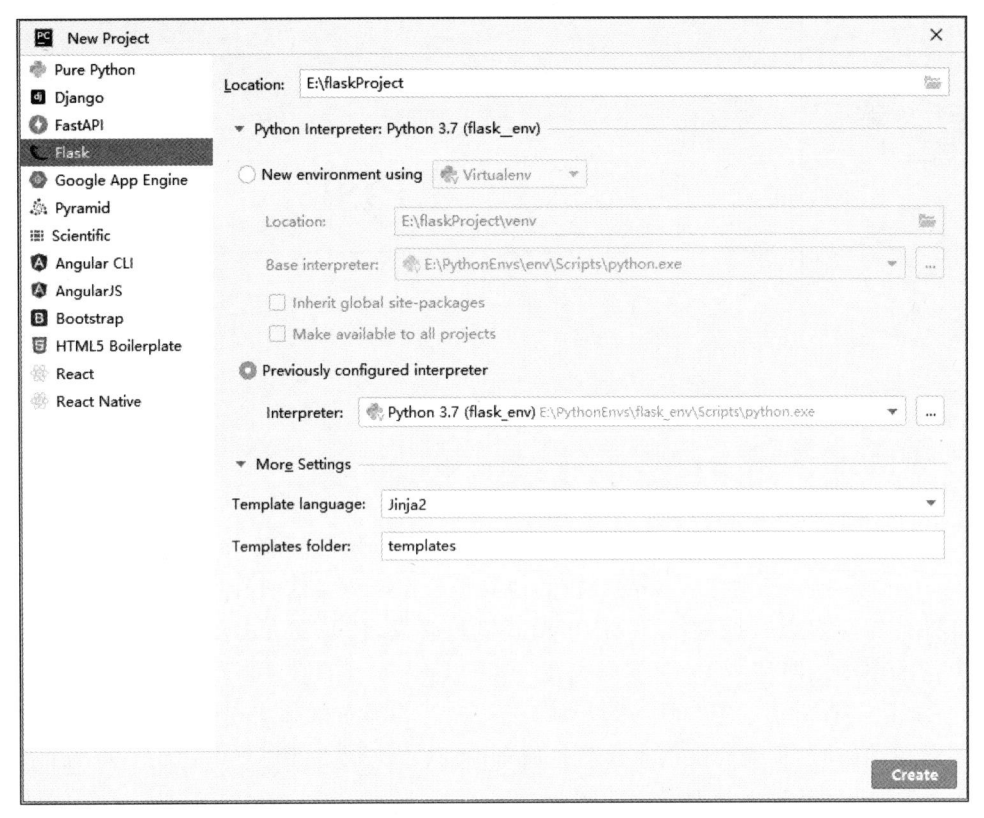

图 3-2　创建 Flask 项目的第 2 步

图 3-3　Flask 的项目结构

static 用于存放 Flask 项目的静态文件,文件夹 templates 用于存放 Flask 项目的模板。

app.py 文件中的代码如下:

```
# 资源包\Code\chapter3\3.3\1\flaskProject\app.py
from flask import Flask
# 创建Flask类的实例对象app,用于处理来自Web服务器的请求数据
app = Flask(__name__)
# 定义路由,用于将视图函数绑定到指定的URL上
@app.route('/')
# 视图函数,用于处理业务逻辑和返回响应内容
def hello_world():
    return 'Hello World!'
if __name__ == '__main__':
    # 在本地服务器上运行应用程序
    app.run()
```

在上述代码中,客户端将请求发送给 Web 服务器之后,Web 服务器会将请求发送给 Flask 实例对象,然后 Flask 实例对象通过路由找到对应的视图函数,在其中处理该请求,并

将响应内容返给 Web 服务器,而 Web 服务器则将响应内容返回客户端,最终在浏览器中显示。

运行上述代码,PyCharm 中的运行结果如图 3-4 所示。

```
FLASK_APP = app.py
FLASK_ENV = development
FLASK_DEBUG = 0
In folder F:/Python全栈开发/flaskProject
F:\Python全栈开发\flaskProject\venv\Scripts\python.exe -m flask run
 * Debug mode: off
WARNING: This is a development server. Do not use it in a production deployment. Use a production WSGI server instead.
 * Running on http://127.0.0.1:5000
Press CTRL+C to quit
```

图 3-4　PyCharm 中的运行结果

此时,打开浏览器并在网址栏输入 http://127.0.0.1:5000,其显示内容如图 3-5 所示。

图 3-5　浏览器中的显示内容

3.4　Flask 实例对象

Flask 实例对象主要用于处理来自 Web 服务器的请求数据,首先通过路由找到对应的视图函数,然后处理请求,并将响应内容返回给 Web 服务器。

可以通过 flask 模块中的 Flask 类创建 Flask 实例对象,其构造函数常用的参数如表 3-1 所示。

表 3-1　Flask 类的构造函数常用参数

参　　数	描　　述
import_name	应用程序所在的模块,其值一般为 __name__
static_url_path	静态文件访问路径
static_folder	静态文件存储的文件夹
template_folder	模板文件存储的文件夹

1. 启动服务器

通过 Flask 实例对象的 run() 方法可以快速地启动本地服务器,其语法格式如下:

```
run(host, port, debug)
```

其中:参数 host 表示要监听的主机名,默认值为 127.0.0.1,如果设置为 0.0.0.0,则可以使服务器被外部访问;参数 port 表示端口号;参数 debug 表示是否开启调试模式。

需要注意的是,该方法启动的是 Flask 自带的应用服务器,无法保证 Flask 项目的性能,而在正式项目中,一般使用 Nginx 和 uWSGI 来部署项目。

2. 调试模式

正常情况下,项目在启动之后,如果对某处代码进行了修改,此时,想要查看最新的运行结果，则需要关闭服务器，然后再次启动才能查看。

而在开发过程中，这无疑是比较浪费时间的，所以 Flask 提供了调试模式，即在启用调试模式后，只需对代码进行修改并保存，程序就会自动重启，此时就可以立即在浏览器中查看最新的运行结果。开启调试模式的具体步骤如下。

（1）调用 Flask 实例对象的 run() 方法，并在其中添加参数 debug，并将其值设置为 True，代码如下：

```python
# 资源包\Code\chapter3\3.4\1\flaskProject\app.py
from flask import Flask
app = Flask(__name__)
@app.route('/')
def hello_world():
    return 'Hello World!'
if __name__ == '__main__':
    # 开启调试模式
    app.run(debug = True)
```

除此之外，还可以调用 Flask 实例对象的 debug 属性，并将其值设置为 True，代码如下：

```python
# 资源包\Code\chapter3\3.4\1\flaskProject\app.py
from flask import Flask
app = Flask(__name__)
@app.route('/')
def hello_world():
    return 'Hello World!'
if __name__ == '__main__':
    # 开启调试模式
    app.debug = True
    app.run()
```

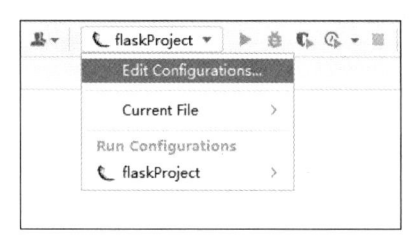

图 3-6　开启调试模式的第 2 步

（2）选择 Edit Configurations，如图 3-6 所示。

（3）勾选 FLASK_DEBUG，然后单击 OK 按钮，即可正常开启调试模式，如图 3-7 所示。

但是，由于 PyCharm 版本的问题，导致部分读者可能无法正常开启调试模式，此时，再次单击 Edit Configurations，并对 flaskProject 界面中 Target 选项的内容进行复制，然后单击左上角的"＋"，选择其中的 Python，如图 3-8 所示。

将刚才复制的内容添加到当前界面的 Script path 选项中，并单击 OK 按钮，如图 3-9 所示。

然后使用新的配置再次运行，如图 3-10 所示。

此时，调试模式已经可以正常开启，如图 3-11 所示。

3．配置项

Flask 的配置项 app.config 是 flask.config.Config 的实例对象，本质上是一个存储配置变量的字典，可以存储各种不同的配置，例如数据库连接字符串、密钥及其他应用程序的

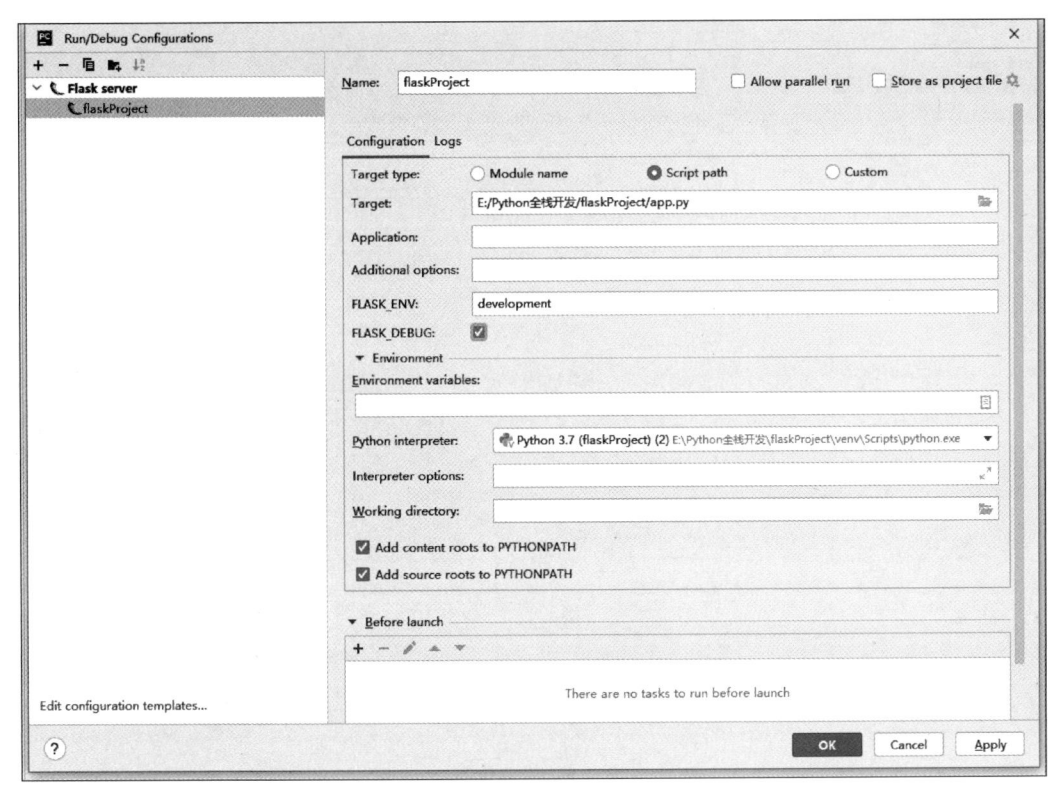

图 3-7　开启调试模式的第 3 步

图 3-8　添加新配置

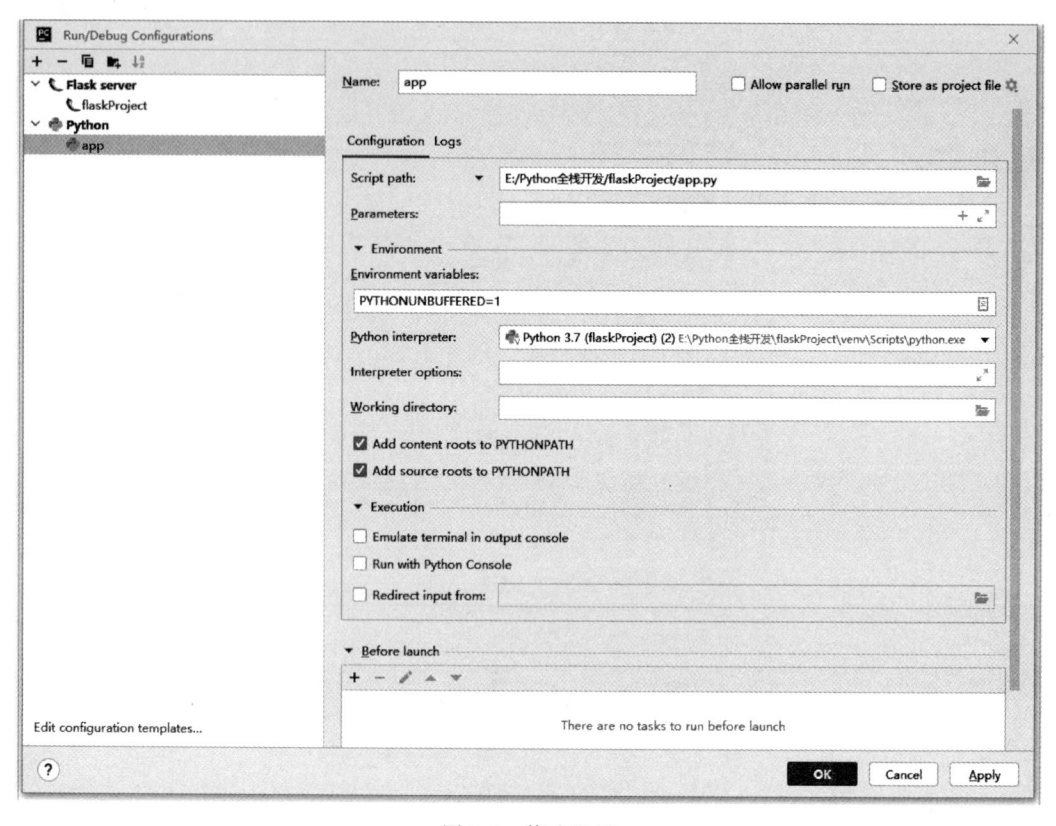

图 3-9　修改配置

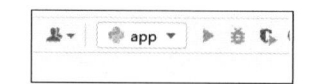

图 3-10　使用 Python 模式运行

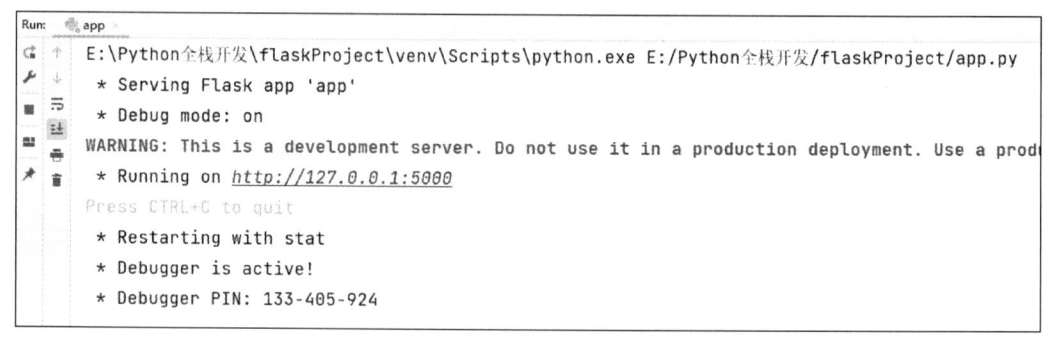

图 3-11　调试模式正常开启

相关设置等。

一般情况下,相关的配置都会在应用程序的配置文件中设置,或者在应用程序启动时从环境变量中加载。

1) 使用 from_object() 方法加载配置对象

下面通过一个示例,演示一下如何加载配置对象。

(1) 创建名为 flaskProject 的 Flask 项目。

（2）在项目的根目录下创建文件 config.py，代码如下：

```
# 资源包\Code\chapter3\3.4\2\flaskProject\config.py
import os
# 开启调试模式
DEBUG = True
# 配置数据库 URI
SQLALCHEMY_DATABASE_URI = 'sqlite://db/database.db'
# 跟踪数据库中数据的修改
SQLALCHEMY_TRACK_MODIFICATIONS = False
# 配置输出 SQL 语句
SQLALCHEMY_ECHO = True
# 创建密钥
SECRET_KEY = os.urandom(24)
```

（3）打开保存在项目根目录下的文件 app.py，代码如下：

```
# 资源包\Code\chapter3\3.4\2\flaskProject\app.py
from flask import Flask
import config
app = Flask(__name__)
app.config.from_object(config)
if __name__ == '__main__':
    app.run()
```

2）使用 from_pyfile() 方法加载配置文件

下面通过一个示例，演示一下如何加载配置文件。

（1）创建名为 flaskProject 的 Flask 项目。

（2）在项目的根目录下创建文件 config.py，代码如下：

```
# 资源包\Code\chapter3\3.4\3\flaskProject\config.py
import os
# 开启调试模式
DEBUG = True
# 配置数据库 URI
SQLALCHEMY_DATABASE_URI = 'sqlite://db/database.db'
# 跟踪数据库中数据的修改
SQLALCHEMY_TRACK_MODIFICATIONS = False
# 配置输出 SQL 语句
SQLALCHEMY_ECHO = True
# 创建密钥
SECRET_KEY = os.urandom(24)
```

（3）打开保存在项目根目录下的文件 app.py，代码如下：

```
# 资源包\Code\chapter3\3.4\3\flaskProject\app.py
from flask import Flask
app = Flask(__name__)
app.config.from_pyfile('config.py')
if __name__ == '__main__':
    app.run()
```

3.5 路由

在 Flask 中,用于处理请求 URL 和视图函数之间关系的程序称为路由。

3.5.1 路由的声明

声明路由的方式有两种。

一是通过 Flask 实例对象所提供的函数装饰器 route 来声明路由,其语法格式如下:

```
route(rule, endpoint)
```

其中:参数 rule 表示请求 URL;参数 endpoint 表示端点,通常用于动态构建 URL,其默认值为视图函数的名称。

下面通过一个示例,演示一下如何声明路由。

(1) 创建名为 flaskProject 的 Flask 项目。

(2) 打开保存在项目根目录下的文件 app.py,代码如下:

```
# 资源包\Code\chapter3\3.5\1\flaskProject\app.py
from flask import Flask
app = Flask(__name__)
@app.route('/oldxia/')
def hello_oldxia():
    return 'www.oldxia.com'
if __name__ == '__main__':
    app.run(debug = True)
```

在上面这段程序中,函数装饰器 route 中的/oldxia/就是请求 URL,而函数装饰器所装饰的函数 hello_oldxia()则是视图函数,主要用于处理业务逻辑和返回响应内容。

(3) 运行上述程序,打开浏览器并在网址栏输入 http://127.0.0.1:5000/oldxia/,其显示内容如图 3-12 所示。

图 3-12 浏览器中的显示内容

二是通过 Flask 实例对象的 add_url_rule()方法添加请求 URL 与视图函数的映射,语法格式如下:

```
add_url_rule(rule, endpoint, view_func)
```

其中:参数 rule 表示请求 URL;参数 endpoint 表示端点,通常用于动态构建 URL,其默认值为视图函数的名称;参数 view_func 表示视图函数的名称。

下面通过一个示例,演示一下如何声明路由。

(1) 创建名为 flaskProject 的 Flask 项目。

(2) 打开保存在项目根目录下的文件 app.py,代码如下:

```
# 资源包\Code\chapter3\3.5\2\flaskProject\app.py
from flask import Flask
```

```
app = Flask(__name__)
def hello_oldxia():
    return 'www.oldxia.com'
app.add_url_rule('/oldxia/', endpoint = 'ho', view_func = hello_oldxia)
if __name__ == '__main__':
    app.run(debug = True)
```

(3) 运行上述程序,打开浏览器并在网址栏输入http://127.0.0.1:5000/oldxia/,其显示内容如图3-13所示。

图3-13 浏览器中的显示内容

3.5.2 唯一规则

唯一规则指的是在路由中定义请求URL时,一定要在该URL的尾部加上斜杠。

虽然,加斜杠和不加斜杠都是同一个URL,但是搜索引擎会将加斜杠和不加斜杠的两个URL视为不同的URL,所以在Flask中唯一规则主要是为了避免搜索引擎对同一页面的重复索引。因为,当一个网站存在大量的重复内容时,势必会影响搜索引擎优化,从而影响网站的权重。简单地讲,也就是同一网站的重复内容越多,搜索引擎对网站的权重评定越低,就越难被浏览器检索到,因此在Flask中保证请求URL的唯一性是非常有必要的。

此外,可以通过在函数装饰器route中添加参数strict_slashes并将其值更改为False来忽略请求URL尾部的斜杠。

下面通过一个示例,演示一下唯一规则的应用。

(1) 创建名为flaskProject的Flask项目。

(2) 打开保存在项目根目录下的文件app.py,代码如下:

```
#资源包\Code\chapter3\3.5\3\flaskProject\app.py
from flask import Flask
app = Flask(__name__)
@app.route('/user/', strict_slashes = False)
def about():
    return 'this is user'
if __name__ == '__main__':
    app.run(debug = True)
```

(3) 此时,打开浏览器,无论是在网址栏中输入http://127.0.0.1:5000/user,或是输入http://127.0.0.1:5000/user/都可以正常得到响应,其显示内容如图3-14所示。

3.5.3 路由分类

根据请求URL的不同,可以将路由分为两种,即静态路由和动态路由。

1. 静态路由

当请求URL不需要发生变化时,可以使用静态路由。

图 3-14 浏览器中的显示内容

下面通过一个示例,演示一下如何配置静态路由。
(1) 创建名为 flaskProject 的 Flask 项目。
(2) 打开保存在项目根目录下的文件 app.py,代码如下:

```python
# 资源包\Code\chapter3\3.5\4\flaskProject\app.py
from flask import Flask
app = Flask(__name__)
@app.route('/')
def hello_world():
    return 'Hello World!'
if __name__ == '__main__':
    app.run()
```

图 3-15 浏览器中的显示内容

(3) 运行上述程序,打开浏览器并在网址栏输入 http://127.0.0.1:5000/,其显示内容如图 3-15 所示。

2. 动态路由

当请求 URL 需要动态变化时,需要使用动态路由,即需要给请求 URL 传递参数,可以通过将请求 URL 中的一部分标记为 < variable_name > 的方式添加参数。

下面通过一个示例,演示一下如何配置动态路由。
(1) 创建名为 flaskProject 的 Flask 项目。
(2) 打开保存在项目根目录下的文件 app.py,代码如下:

```python
# 资源包\Code\chapter3\3.5\5\flaskProject\app.py
from flask import Flask
app = Flask(__name__)
@app.route('/article/<id>/')
def article_list(id):
    return f'您所请求的文章 id 是【{id}】'
if __name__ == '__main__':
    app.run(debug=True)
```

在上面的代码中 <id> 就是请求 URL 中的参数,需要注意的是,如果请求 URL 中包含参数,则视图函数中必须传递该参数。

(3) 运行上述程序,打开浏览器并在网址栏输入 http://127.0.0.1:5000/article/1001/,其显示内容如图 3-16 所示。

图 3-16　浏览器中的显示内容

3.5.4　动态构建请求 URL

动态构建请求 URL 有两大优点：一是当视图函数所对应的请求 URL 被大量修改时，无须手动去更改其他地方已经使用该视图函数所对应的请求 URL；二是动态构建请求 URL 会自动处理特殊的字符，而不需要手动进行处理。

可以通过 flask 模块中的 url_for() 函数动态地构建指定视图函数的请求 URL，其语法格式如下：

```
url_for(endpoint, values)
```

其中，参数 endpoint 表示端点，默认值为视图函数的名称。需要注意的是，如果需要构建 URL 的视图函数使用参数 endpoint 指定了名称，则必须使用该名称，否则会报错；当没有使用该参数指定名称时，可以直接使用视图函数的名称。参数 values 表示请求 URL 中的参数，如果出现未知参数，则会添加到请求 URL 中作为查询参数。

下面通过一个示例，演示一下如何动态地构建请求 URL。

（1）创建名为 flaskProject 的 Flask 项目。

（2）打开保存在项目根目录下的文件 app.py，代码如下：

```
#资源包\Code\chapter3\3.5\6\flaskProject\app.py
from flask import Flask, url_for
app = Flask(__name__)
@app.route('/user/')
def user():
    return f'this is index'
@app.route('/about/<id>/')
def article(id):
    return f'this is article id is 【{id}】'
@app.route('/ufi/')
def url_for_index():
    return url_for('user')
@app.route('/ufa1/')
def url_for_article_one():
    return url_for('article', id=1001)
@app.route('/ufa2/')
def url_for_article_two():
    return url_for('article', id=1001, name='xzd')
if __name__ == '__main__':
    app.run(debug=True)
```

（3）运行上述程序，打开浏览器并在网址栏输入 http://127.0.0.1:5000/ufi/，其显示内容如图 3-17 所示。

(4)在浏览器中的网址栏输入http://127.0.0.1:5000/ufa1/,其显示内容如图3-18所示。

图3-17 浏览器中的显示内容

图3-18 浏览器中的显示内容

(5)在浏览器中的网址栏输入http://127.0.0.1:5000/ufa2/,其显示内容如图3-19所示。

图3-19 浏览器中的显示内容

3.5.5 HTTP请求

当在浏览器中的网址栏输入URL并按Enter键时,客户端便会以某种请求方法发起一个HTTP请求,常用的请求方法如表3-2所示。

表3-2 常用的请求方法

请 求 方 法	描 述
GET	请求指定的页面信息,并返回实体主体
HEAD	类似于GET请求,只不过返回的响应中没有具体的内容,用于获取报头
POST	向指定资源提交数据并处理请求,数据被包含在请求体中
PUT	从客户端向服务器端传送的数据取代指定文档中的内容
DELETE	请求服务器删除指定的页面
CONNECT	HTTP1.1协议中预留给能够将连接改为管道方式的代理服务器
OPTIONS	允许客户端查看服务器的性能
TRACE	回显服务器收到的请求,主要用于测试或诊断
PATCH	PUT方法的补充,用来对已知资源进行局部更新

在默认情况下,Flask项目中的路由只能响应GET请求,但可以通过函数装饰器route中的methods参数来指定其他HTTP请求方法。

此外,还可以通过flask模块中request对象的相关属性来获取GET请求或POST请求中参数所对应的值,常用的属性如表3-3所示。

表3-3 request对象的常用属性

属 性	描 述
method	获取请求方法
args	获取查询字符串中的内容

续表

属　　性	描　　述
form	获取表单中的内容
cookies	获取 Cookie 的信息
headers	获取请求头的信息
files	获取上传或下载的文件信息

下面通过一个示例，演示一下如何发起 GET 和 POST 请求。

（1）创建名为 flaskProject 的 Flask 项目。

（2）打开保存在项目根目录下的文件 app.py，代码如下：

```
# 资源包\Code\chapter3\3.5\7\flaskProject\app.py
from flask import Flask, request
app = Flask(__name__)
@app.route('/')
def index():
    return '''
    <!doctype html>
    <html lang="en">
    <head>
        <meta charset="UTF-8">
        <title>Document</title>
    </head>
    <body>
        <form action="/login/" method="post">
            名称：<input type="text" name="username"><br><br>
            密码：<input type="password" name="password"><br><br>
            <input type="submit" value="提交">
        </form>
    </body>
    </html>
    '''
@app.route('/login/', methods=['GET', 'POST'])
def login():
    if request.method == 'GET':
        return f'GET 请求的全部参数{request.args},其中参数 username 的值为{request.args.get("username")}'
    elif request.method == 'POST':
        return f'POST 请求的全部参数{request.form},其中参数 password 的值为{request.form.get("password")}'
if __name__ == '__main__':
    app.run(debug=True)
```

（3）运行上述程序，打开浏览器并在网址栏输入 http://127.0.0.1:5000/login/?username=xzd&age=36，其显示内容如图 3-20 所示。

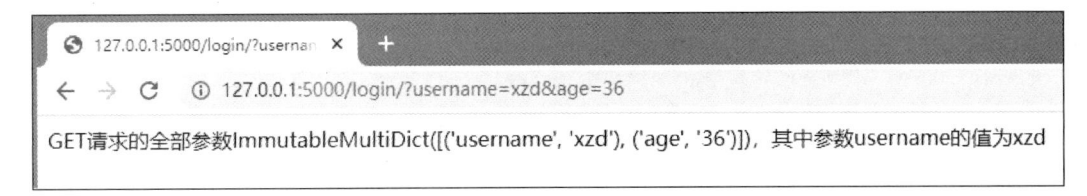

图 3-20　浏览器中的显示内容

(4) 在浏览器中的网址栏输入 http://127.0.0.1:5000/，并进行登录，其显示内容如图 3-21 所示。

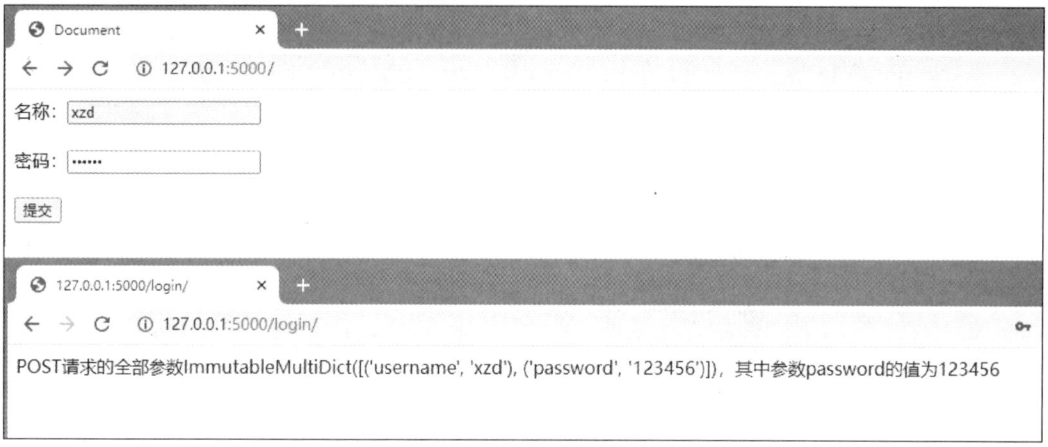

图 3-21　浏览器中的显示内容

3.5.6　路由转换器

路由转换器主要用于对请求 URL 的参数类型进行限制，其可以分为内置转换器和自定义转换器。

1. 内置转换器

通过<converter:variable_name>的方式添加内置转换器，其中 converter 表示限制的规则，包括 string（默认转换器，接受任何不包含斜杠的文本）、int（接受正整数）、float（接受正浮点数）、path（类似 string，但可以接受包含斜杠的文本）、uuid（接受 uuid 字符串）和 any（匹配多个路径）。

下面通过一个示例，演示一下如何使用内置转换器。

(1) 创建名为 flaskProject 的 Flask 项目。

(2) 打开保存在项目根目录下的文件 app.py，代码如下：

```
# 资源包\Code\chapter3\3.5\8\flaskProject\app.py
from flask import Flask
app = Flask(__name__)
@app.route('/article_string/<string:id>/')
def article_string_list(id):
    return f'您所请求的文章id是【{id}】'
@app.route('/article_int/<int:id>/')
def article_int_list(id):
    return f'您所请求的文章id是【{id}】'
@app.route('/article_float/<float:id>/')
def article_float_list(id):
    return f'您所请求的文章id是【{id}】'
@app.route('/article_path/<path:id>/')
def article_path_list(id):
    return f'您所请求的文章id是【{id}】'
@app.route('/article_uuid/<uuid:id>/')
def article_uuid_list(id):
```

```
    return f'您所请求的文章id是【{id}】'
@app.route('/article_any/<any(news,blog):url_path>/<id>/')
def article_any_list(url_path, id):
    if url_path == "news":
        return f'您所访问的是【新闻页面】,文章id是【{id}】'
    else:
        return f'您所访问的是【博客页面】,文章id是【{id}】'
if __name__ == '__main__':
    app.run(debug = True)
```

(3) 运行上述程序,打开浏览器并在网址栏输入 http://127.0.0.1:5000/article_string/first/,其显示内容如图 3-22 所示。

(4) 在浏览器中的网址栏输入 http://127.0.0.1:5000/article_int/1001/,其显示内容如图 3-23 所示。

图 3-22　浏览器中的显示内容

图 3-23　浏览器中的显示内容

(5) 在浏览器中的网址栏输入 http://127.0.0.1:5000/article_float/1001.01/,其显示内容如图 3-24 所示。

(6) 在浏览器中的网址栏输入 http://127.0.0.1:5000/article_path/first/second/,其显示内容如图 3-25 所示。

图 3-24　浏览器中的显示内容

图 3-25　浏览器中的显示内容

(7) 在浏览器中的网址栏输入 http://127.0.0.1:5000/article_uuid/123e4567-e89b-4123-b567-0987654321ab/,其显示内容如图 3-26 所示。

图 3-26　浏览器中的显示内容

(8) 在浏览器中的网址栏输入 http://127.0.0.1:5000/article_any/news/1001/,其显示内容如图 3-27 所示。

2. 自定义转换器

创建自定义转换器分为两步。

图 3-27 浏览器中的显示内容

（1）创建转换器类，并在该类中实现所需请求 URL 参数的规则，需要注意的是，该类需要继承自 werkzeug.routing 模块中的 BaseConverter 类，其常用的属性和方法如表 3-4 所示。

表 3-4　BaseConverter 的常用属性和方法

属　　性	描　　述
regex	匹配请求 URL 参数的正则表达式
方法	描述
to_python()	当匹配到请求 URL 的参数后，该方法会将其返回值传递到该请求 URL 所对应的视图函数中
to_url()	当其他视图函数使用 url_for() 函数时，该方法会对传入的请求 URL 参数进行处理并返回

（2）将自定义的转换器添加至转换器字典中。

下面通过一个示例，演示一下如何自定义转换器。

（1）创建名为 flaskProject 的 Flask 项目。

（2）打开保存在项目根目录下的文件 app.py，代码如下：

```python
#资源包\Code\chapter3\3.5\9\flaskProject\app.py
from flask import Flask, url_for
from werkzeug.routing import BaseConverter
app = Flask(__name__)
#自定义转换器类
class TelephoneConverter(BaseConverter):
    regex = r'1[3456789]\d{9}'
class AListConverter(BaseConverter):
    def to_python(self, value):
        print("to_python方法被调用")
        return f'{value} - to_python()'
    def to_url(self, value):
        print("to_url方法被调用")
        return ' - '.join(value)
#将自定义的转换器添加至转换器字典中
app.url_map.converters['tel'] = TelephoneConverter
app.url_map.converters['alist'] = AListConverter
@app.route('/')
def index():
    return url_for('parts', parm=['python','linux'])
@app.route('/user/<tel:my_tel>/')
def user(my_tel):
    return f'手机号码为{my_tel}'
@app.route('/parts/<alist:parm>/')
def parts(parm):
```

```
        return f'{parm}'
if __name__ == '__main__':
    app.run(debug = True)
```

(3) 运行上述程序,打开浏览器并在网址栏输入http://127.0.0.1:5000/,其显示内容如图3-28所示。

(4) 在浏览器中的网址栏输入http://127.0.0.1:5000/user/13309861086/,其显示内容如图3-29所示。

图3-28　浏览器中的显示内容

图3-29　浏览器中的显示内容

(5) 在浏览器中的网址栏输入http://127.0.0.1:5000/parts/xzd/,其显示内容如图3-30所示。

图3-30　浏览器中的显示内容

3.5.7　重定向

重定向分为永久性重定向和暂时性重定向,在页面上体现的操作就是浏览器会从一个页面自动跳转到另外一个页面,例如用户访问了一个需要权限的页面,但是该用户当前并没有登录,因此需要将该用户重定向至登录页面。

在Flask中,可以通过flask模块中的redirect()函数实现重定向,其语法格式如下:

```
redirect(location, code)
```

其中:参数location表示重定向的URL;参数code表示重定向的响应状态码,默认值为302,即暂时性重定向,也可修改为301,实现永久性重定向。

下面通过一个示例,演示一下如何进行重定向。

(1) 创建名为flaskProject的Flask项目。

(2) 打开保存在项目根目录下的文件app.py,代码如下:

```
# 资源包\Code\chapter3\3.5\10\flaskProject\app.py
from flask import Flask, url_for, redirect, request
app = Flask(__name__)
@app.route('/')
def index():
    # 判断是否登录
    if request.args.get('username') is None:
```

```
        return redirect(url_for('login'))
    #登录成功
    username = request.args.get('username')
    return f'登录成功,欢迎用户:{username}'
@app.route('/login/')
def login():
    return '先登录!'
if __name__ == '__main__':
    app.run(debug = True)
```

(3) 运行上述程序,打开浏览器并在网址栏输入 http://127.0.0.1:5000/login/,模拟未登录的场景,其显示内容如图 3-31 所示。

(4) 在浏览器中的网址栏输入 http://127.0.0.1:5000/?username=xzd,模拟登录成功时的场景,其显示内容如图 3-32 所示。

图 3-31　浏览器中的显示内容 　　　　图 3-32　浏览器中的显示内容

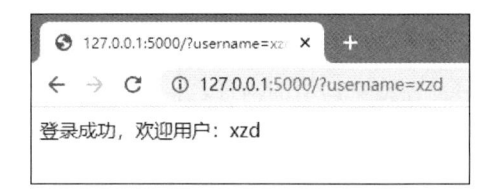

3.5.8　视图函数的返回值

在 Flask 中,HTTP 响应主要通过 Response 对象实现。

HTTP 响应报文中的绝大部分内容均由服务器处理,所以一般情况下,开发人员只需负责返回主体内容,而视图函数的返回值则构成了 HTTP 响应报文的主体内容。

Response 对象的创建可以通过 flask 模块中的 Response 类或 make_response() 函数进行实现。

下面通过一个示例,演示一下如何创建 Response 对象。

(1) 创建名为 flaskProject 的 Flask 项目。

(2) 打开保存在项目根目录下的文件 app.py,代码如下:

```
#资源包\Code\chapter3\3.5\11\flaskProject\app.py
from flask import Flask, make_response, Response
app = Flask(__name__)
@app.route('/response/')
def r():
    response = Response('Response', 200, {'name': 'xzd'})
    return response
@app.route('/make_response/')
def m_r():
    response = make_response('make_response', 200, {'age': 36})
    return response
if __name__ == '__main__':
    app.run(debug = True)
```

(3) 运行上述程序,打开浏览器并在网址栏输入 http://127.0.0.1:5000/response/,其显示内容如图3-33所示。

(4) 在浏览器中的网址栏输入 http://127.0.0.1:5000/make_response/，其显示内容如图3-34所示。

图3-33　浏览器中的显示内容

图3-34　浏览器中的显示内容

此外，视图函数虽然可以返回众多不同类型的数据，例如字符串、字典、列表和元组等，但本质上，这些数据最终都将被封装为 Response 对象。

下面通过一个示例，演示一下视图函数的返回值。

(1) 创建名为 flaskProject 的 Flask 项目。

(2) 打开保存在项目根目录下的文件 app.py，代码如下：

```python
# 资源包\Code\chapter3\3.5\12\flaskProject\app.py
from flask import Flask
app = Flask(__name__)
@app.route('/str/')
def r_str():
    return 'hello Flask'
@app.route('/dict/')
def r_dict():
    return {'tel': '13309861086'}
@app.route('/tuple/')
def r_tuple():
    # 响应体为 hello Flask；响应状态码为200；响应头为 name:oldxia
    return ('hello Flask', 200, {'name': 'oldxia'})
@app.route('/list/')
def r_list():
    return ['xzd', '36']
if __name__ == '__main__':
    app.run(debug = True)
```

(3) 运行上述程序，打开浏览器并在网址栏输入 http://127.0.0.1:5000/str/，其显示内容如图3-35所示。

(4) 在浏览器中的网址栏输入 http://127.0.0.1:5000/dict/，其显示内容如图3-36所示。

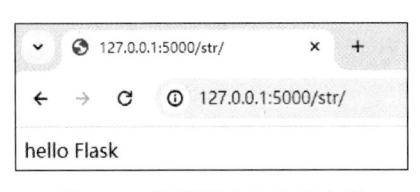

图3-35　浏览器中的显示内容

图3-36　浏览器中的显示内容

(5) 在浏览器中的网址栏输入 http://127.0.0.1:5000/tuple/，其显示内容如图3-37

(6) 在浏览器中的网址栏输入 http://127.0.0.1:5000/list/，其显示内容如图 3-38 所示。

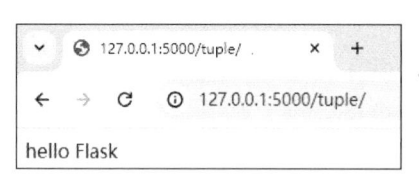

图 3-37　浏览器中的显示内容 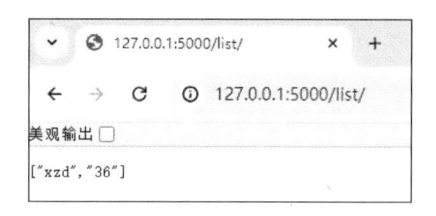　　　　图 3-38　浏览器中的显示内容

3.5.9　自定义视图函数装饰器

在 Flask 中，视图函数可以通过函数装饰器赋予额外的功能，例如之前所学习的函数装饰器 route()，但是，在一些特殊的情况下，往往需要自定义视图函数装饰器，以满足开发的需求。

这里需要注意的是，在自定义视图函数装饰器中，务必使用 functools 模块中的 wraps() 函数装饰其内部的函数，这可以确保自定义函数装饰器不会对被装饰的函数造成影响，即在扩展函数功能的同时，保留原有函数的各种属性。

下面通过一个示例，演示一下如何自定义视图函数装饰器。

(1) 创建名为 flaskProject 的 Flask 项目。

(2) 打开保存在项目根目录下的文件 app.py，代码如下：

```python
# 资源包\Code\chapter3\3.5\13\flaskProject\app.py
import functools
from flask import Flask, request
app = Flask(__name__)
# 自定义的函数装饰器
def login_required(func):
    @functools.wraps(func)
    def wraps(*args, **kwargs):
        # func 表示被自定义函数装饰器所装饰的视图函数
        print(func)
        username = request.args.get('username')
        if username and username == 'oldxia':
            return func(*args, **kwargs)
        else:
            return '<h1 style="color: red">先登录</h1>'
    return wraps
@app.route('/course/')
@login_required
def course():
    return '欢迎进入选课中心'
if __name__ == '__main__':
    app.run(debug=True)
```

(3) 运行上述程序，打开浏览器并在网址栏输入 http://127.0.0.1:5000/course/，其显示内容如图 3-39 所示。

(4) 在浏览器中的网址栏输入 http://127.0.0.1:5000/course/?username=oldxia，模拟登录成功时的场景，其显示内容如图 3-40 所示。

图 3-39　浏览器中的显示内容

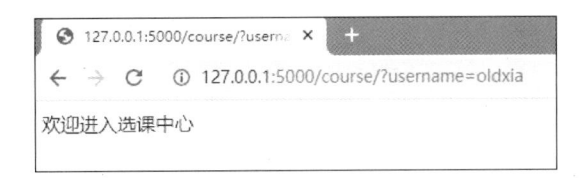

图 3-40　浏览器中的显示内容

3.6　模板

在大型项目中，后端主要负责业务逻辑及数据访问，而前端主要负责表现及交互逻辑，而如果将业务逻辑和表现内容放在一起，则会大幅增加代码的复杂度和维护成本。

所以，在实际开发过程中，为有效提升开发效率，需要使前后端分离，即将表现内容交给模板引擎，而模板引擎通过渲染，使用真实值替换网页模板中的变量，进而生成一个标准的 HTML 文档。

在 Flask 中，通常使用 Jinja2 模板引擎实现复杂的页面渲染。

3.6.1　渲染模板

通过 flask 模块中的 render_template() 函数对模板进行渲染，需要注意的是，该函数的返回值为字符串类型，其语法格式如下：

```
render_template(template_name, context)
```

其中，参数 template_name 表示模板的文件名，参数 context 表示模板中的变量。

下面通过一个示例，演示一下如何渲染模板。

(1) 创建名为 flaskProject 的 Flask 项目。

(2) 打开保存在项目根目录下的文件 app.py，代码如下：

```
#资源包\Code\chapter3\3.6\1\flaskProject\app.py
from flask import Flask, render_template
app = Flask(__name__)
@app.route('/')
def index():
    return render_template('index.html')
if __name__ == '__main__':
    app.run()
```

(3) 在项目的根目录下的文件夹 templates 中创建模板文件 index.html，代码如下：

```
#资源包\Code\chapter3\3.6\1\flaskProject\templates\app.py
<!DOCTYPE html>
<html lang="en">
    <head>
        <meta charset="UTF-8">
```

```html
        <meta name="viewport" content="width=device-width, initial-scale=1.0">
        <title>Document</title>
    </head>
    <body>
        <h1>这是模板文件</h1>
    </body>
</html>
```

（4）运行上述程序，打开浏览器并在网址栏输入 http://127.0.0.1:5000/，其显示内容如图 3-41 所示。

图 3-41　浏览器中的显示内容

3.6.2　模板位置

在默认情况下，在渲染模板时，模板文件会存放在项目根目录下的文件夹 templates 之中。

此外，可以在创建 Flask 实例对象时，通过添加参数 template_folder 来指定模板的位置。

下面通过一个示例，演示一下如何设置模板位置。

（1）创建名为 flaskProject 的 Flask 项目。

（2）打开保存在项目根目录下的文件 app.py，代码如下：

```python
# 资源包\Code\chapter3\3.6\2\flaskProject\app.py
from flask import Flask, render_template
app = Flask(__name__, template_folder='d:/templates')
@app.route('/')
def index():
    return render_template('index.html')
if __name__ == '__main__':
    app.run(debug=True)
```

（3）在 D 盘的根目录下创建文件夹 templates，并在其中创建模板文件 index.html，代码如下：

```html
# 资源包\Code\chapter3\3.6\2\templates\index.html
<!DOCTYPE html>
<html lang="en">
    <head>
        <meta charset="UTF-8">
        <meta name="viewport" content="width=device-width, initial-scale=1.0">
        <title>Document</title>
```

```
        </head>
        <body>
            <h1>这是更改位置之后的模板文件</h1>
        </body>
</html>
```

(4) 运行上述程序,打开浏览器并在网址栏输入 http://127.0.0.1:5000/,其显示内容如图 3-42 所示。

图 3-42 浏览器中的显示内容

3.6.3 模板变量

在模板中可以通过标签{{variable_name}}来获取 render_template()函数所传递的参数,并在渲染模板时将其解析成对应的值。

此外,如果需要传递多个参数,则可以首先通过定义字典将多个参数存放至该字典中,然后在 render_template()函数中将该字典作为关键字参数传入。

下面通过一个示例,演示一下如何使用模板变量。

(1) 创建名为 flaskProject 的 Flask 项目。

(2) 打开保存在项目根目录下的文件 app.py,代码如下:

```
# 资源包\Code\chapter3\3.6\3\flaskProject\app.py
from flask import Flask, render_template
app = Flask(__name__)
@app.route('/')
def index():
    return render_template('index.html', st='oldxia', lt=[1, 2, 3], dt={'name': 'xzd', 'age': 36})
@app.route('/multi/')
def multi():
    context = {
        'st': 'oldxia',
        'lt': [1, 2, 3, 4, 5, 6],
        'dt': {'name': 'xzd', 'age': 36}
    }
    return render_template('index.html', **context)
if __name__ == '__main__':
    app.run(debug=True)
```

(3) 在项目的根目录下的文件夹 templates 中创建模板文件 index.html,代码如下:

```
# 资源包\Code\chapter3\3.6\3\flaskProject\templates\index.html
<!DOCTYPE html>
```

```html
<html lang = "en">
    <head>
        <meta charset = "UTF-8">
        <meta name = "viewport" content = "width = device-width, initial-scale = 1.0">
        <title>Document</title>
    </head>
    <body>
    <h1>模板文件</h1>
    <h2>变量 st-{{ st }}</h2>
    <h2>变量 lt-{{ lt }}</h2>
    <h2>变量 dt-{{ dt }}</h2>
    </body>
</html>
```

（4）运行上述程序，打开浏览器并在网址栏输入 http://127.0.0.1:5000/，其显示内容如图 3-43 所示。

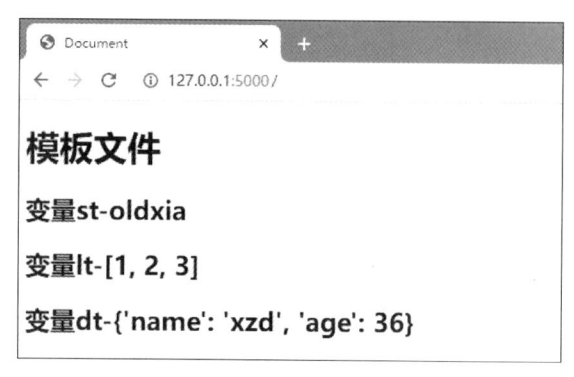

图 3-43　浏览器中的显示内容

（5）在浏览器中的网址栏输入 http://127.0.0.1:5000/multi/，其显示内容如图 3-44 所示。

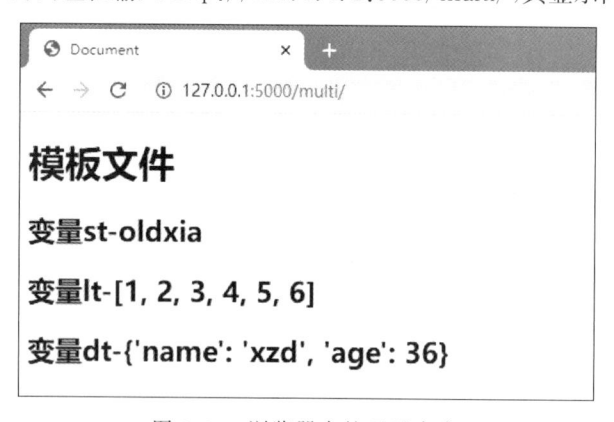

图 3-44　浏览器中的显示内容

3.6.4　模板中动态构建请求 URL

在模板文件中同样可以使用 url_for() 函数动态地构建请求 URL。

下面通过一个示例，演示一下如何在模板中动态地构建请求 URL。

（1）创建名为 flaskProject 的 Flask 项目。

（2）打开保存在项目根目录下的文件 app.py，代码如下：

```
# 资源包\Code\chapter3\3.6\4\flaskProject\app.py
from flask import Flask, render_template
app = Flask(__name__)
@app.route('/')
def index():
    return render_template('index.html')
@app.route('/login/<name>/')
def login(name):
    return render_template('login.html')
if __name__ == '__main__':
    app.run(debug=True)
```

（3）在项目的根目录下的文件夹 templates 中创建模板文件 index.html，代码如下：

```
# 资源包\Code\chapter3\3.6\4\flaskProject\templates\index.html
<!DOCTYPE html>
<html lang="en">
    <head>
        <meta charset="UTF-8">
        <meta name="viewport" content="width=device-width, initial-scale=1.0">
        <title>Document</title>
    </head>
    <body>
        <h1><a href="{{ url_for('login', name='xzd', age='36') }}">单击此处进行登录</a></h1>
    </body>
</html>
```

（4）在项目的根目录下的文件夹 templates 中创建模板文件 login.html，代码如下：

```
# 资源包\Code\chapter3\3.6\4\flaskProject\templates\login.html
<!DOCTYPE html>
<html lang="en">
    <head>
        <meta charset="UTF-8">
        <meta name="viewport" content="width=device-width, initial-scale=1.0">
        <title>Document</title>
    </head>
    <body>
        <h1>登录成功后的页面</h1>
    </body>
</html>
```

（5）运行上述程序，打开浏览器并在网址栏输入 http://127.0.0.1:5000/，其显示内容如图 3-45 所示。

（6）单击"单击此处进行登录"按钮，页面将跳转至动态构建的请求 URL，其显示内容如图 3-46 所示。

3.6.5　模板中的过滤器

在模板中，过滤器相当于一个特殊的函数，主要用于修改和过滤变量的值。过滤器通过管道符"|"实现，其使用方式如下：

图 3-45　浏览器中的显示内容　　　　　　　图 3-46　浏览器中的显示内容

```
{{ variable|filter }}
```

其中，variable 表示模板变量，filter 表示过滤器的名称。

1. 内置过滤器

在 Jinja2 中，内置了许多过滤器，如表 3-5 所示。

表 3-5　常用的内置过滤器

过滤器	描　　述
abs	用于返回一个数值的绝对值
default	当参数 boolean 的值为 False 且当前的模板变量没有传递值时，使用参数 default_value 的值进行代替；当参数 boolean 的值为 True 时，将通过判断模板变量的布尔值来决定是否使用参数 default_value 的值进行代替，即当模板变量的布尔值为 True 时，使用参数 default_value 的值进行代替，反之则不使用。此外，当过滤器的参数 boolean 的值为 True 时，还可以使用 or 进行替换
escape	用于将字符串中的特殊符号（如<、>、& 等）进行转义
safe	用于关闭字符串的自动转义
first	返回序列中的第 1 个元素
last	返回序列中的最后一个元素
length	返回序列或字典的长度
join	将序列用参数 d 的值拼接成字符串
int	将变量的值转换为整数
float	将变量的值转换为浮点数
string	将变量的值转换为字符串
lower	将字符串转换为小写
upper	将字符串转换为大写
replace	将字符串中的参数 old 所指定的字符串替换成参数 new 所指定的字符串
truncate	截取参数 length 所指定长度的字符串。可分为 3 种情况：一是当字符串长度小于参数 length 的值加上参数 leeway（默认值为 5 个字符）的值时，将直接返回传入的字符串；二是当字符串长度大于参数 length 的值加上参数 leeway 的值且当参数 killwords 为 True 时，先截取参数 length 所表示的指定长度减去参数 end 所表示的结束符长度的字符串，最后将该字符串与参数 end 所表示的结束符号进行连接并返回；三是当字符串长度大于参数 length 的值加上参数 leeway 的值且当参数 killwords 为 False 时，先截取参数 length 所表示的指定长度减去参数 end 所表示的结束符长度的字符串，然后调用该字符串的 rsplit() 函数以空格为分隔符进行 1 次分割，并获取分割后的第一部分字符串，最后将该字符串与参数 end 所表示的结束符号进行连接并返回

过 滤 器	描 述
striptags	用于删除字符串中所有的 HTML 标签。此外，如果出现多个空格，则将替换成一个空格
wordcount	用于计算字符串中单词的个数
trim	截取字符串前面和后面的空白字符

下面通过一个示例，演示一下如何使用模板中的内置过滤器。

（1）创建名为 flaskProject 的 Flask 项目。

（2）打开保存在项目根目录下的文件 app.py，代码如下：

```python
# 资源包\Code\chapter3\3.6\5\flaskProject\app.py
from flask import Flask, render_template
app = Flask(__name__)
@app.route('/')
def index():
    context = {
        'position': -9,
        'text': None,
        'signature': '<script>alert("老夏学院")</script>',
        'persons': ['Python', 'PHP', 'JavaScript'],
        'article': 'hello www.oldxia.com',
        'info': {'name': 'xzd', 'age': 36},
        'books': ['Python全栈开发——基础入门', 'Python全栈开发——高阶编程', 'Python全栈开发——数据分析', ],
        'age': "36",
        'content': 'How are you ?',
        'tag': ' this is Flask ',
    }
    return render_template('index.html', **context)
if __name__ == '__main__':
    app.run(debug=True)
```

（3）在项目的根目录下的文件夹 templates 中创建模板文件 index.html，代码如下：

```html
# 资源包\Code\chapter3\3.6\5\flaskProject\templates\index.html
<!DOCTYPE html>
<html lang="en">
    <head>
        <meta charset="UTF-8">
        <meta name="viewport" content="width=device-width, initial-scale=1.0">
        <title>Document</title>
    </head>
    <body>
        <p>过滤器 abs{{ position|abs }}</p>
        <p>过滤器 default{{ text|default(default_value='此人很懒,没有留下任何说明!') }}</p>
        <p>过滤器 default{{ text|default(default_value='此人很懒,没有留下任何说明!', boolean=True) }}</p>
        <p>过滤器 default{{ text or '此人很懒,没有留下任何说明!' }}</p>
        <p>过滤器 escape{{ signature|escape }}</p>
        <p>过滤器 safe{{ signature|safe }}</p>
        <p>过滤器 first{{ persons|first }}</p>
```

```html
        <p>过滤器 last【{{ persons|last }}】</p>
        <p>过滤器 length【{{ article|length }}】</p>
        <p>过滤器 length【{{ info|length }}】</p>
        <p>过滤器 join【{{ books|join('-') }}】</p>
        <p>过滤器 int【{{ age|int }}】</p>
        <p>过滤器 float【{{ age|float }}】</p>
        <p>过滤器 string【{{ age|string }}】</p>
        <p>过滤器 lower【{{ content|lower }}】</p>
        <p>过滤器 upper【{{ content|upper }}】</p>
        <p>过滤器 replace【{{ tag|replace('Flask','Python') }}】</p>
        <p>过滤器 truncate【{{ content|truncate(length=6,killwords=True) }}】</p>
        <p>过滤器 truncate【{{ content|truncate(length=6,killwords=False) }}】</p>
        <p>过滤器 striptags【{{ signature|striptags }}】</p>
        <p>过滤器 wordcount【{{ tag|wordcount }}】</p>
        <p>过滤器 trim【{{ tag|trim }}】</p>
    </body>
</html>
```

（4）运行上述程序，打开浏览器并在网址栏输入 http://127.0.0.1:5000/，其显示内容如图 3-47 所示。

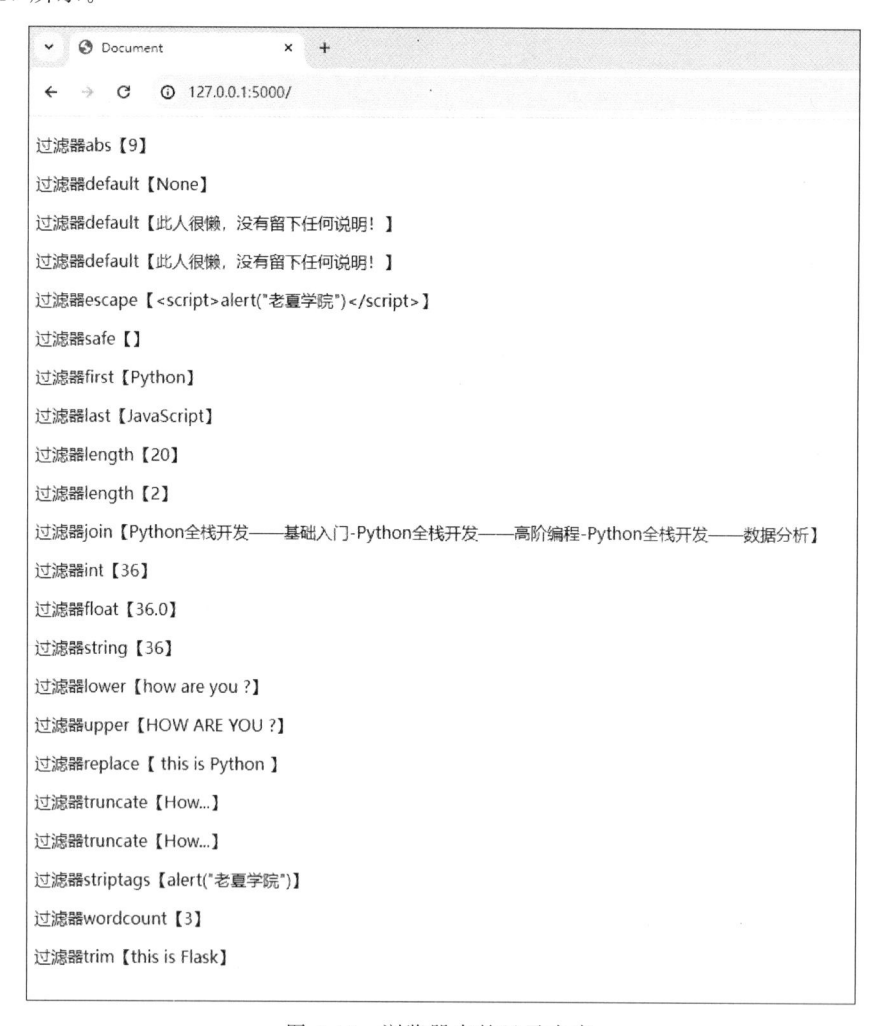

图 3-47　浏览器中的显示内容

2. 自定义过滤器

在实际开发过程中,内置过滤器往往无法满足一些特殊的需求,因此 Jinjia2 还支持自定义过滤器。

自定义过滤器的本质就是自定义一个函数。在模板中,可以通过以下两种方式自定义过滤器。

一是通过 Flask 实例对象的 add_template_filter() 方法实现自定义过滤器,其语法格式如下:

```
add_template_filter(f, name)
```

其中,参数 f 表示自定义过滤器函数,参数 name 表示自定义过滤器的名称。

二是通过 Flask 实例对象所提供的函数装饰器 template_filter 来实现自定义过滤器。

下面通过一个示例,演示一下如何使用模板中的自定义过滤器。

(1) 创建名为 flaskProject 的 Flask 项目。

(2) 打开保存在项目根目录下的文件 app.py,代码如下:

```python
# 资源包\Code\chapter3\3.6\6\flaskProject\app.py
from flask import Flask, render_template
app = Flask(__name__)
@app.route('/')
def index():
    return render_template('index.html', lt=[1, 2, 3, 4, 5, 6, 7, 8])
# 第1种方式
def do_listreverse(li):
    temp_li = list(li)
    temp_li.reverse()
    return temp_li
app.add_template_filter(do_listreverse, 'lireverse1')
# 第2种方式
@app.template_filter('lireverse2')
def do_listreverse(li):
    temp_li = list(li)
    temp_li.reverse()
    return temp_li
if __name__ == '__main__':
    app.run(debug=True)
```

(3) 在项目的根目录下的文件夹 templates 中创建模板文件 index.html,代码如下:

```html
# 资源包\Code\chapter3\3.6\6\flaskProject\templates\index.html
<!DOCTYPE html>
<html lang="en">
    <head>
        <meta charset="UTF-8">
        <meta name="viewport" content="width=device-width, initial-scale=1.0">
        <title>Document</title>
    </head>
    <body>
        <p>原列表:{{ lt }}</p>
        <p>使用自定义过滤器 lireverse1 反转之后的列表:{{ lt|lireverse1 }}</p>
        <p>使用自定义过滤器 lireverse2 反转之后的列表:{{ lt|lireverse2 }}</p>
    </body>
</html>
```

（4）运行上述程序，打开浏览器并在网址栏输入 http://127.0.0.1:5000/，其显示内容如图 3-48 所示。

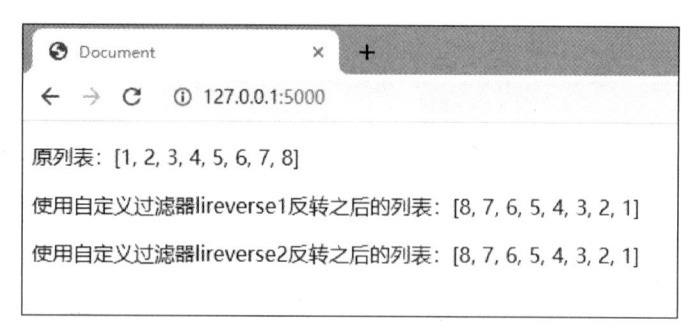

图 3-48　浏览器中的显示内容

3.6.6　模板中的控制结构

Jinja2 提供了多种控制结构，用于改变模板的渲染流程。

1．选择结构

1）单分支选择结构

使用标签{% if %}…{% endif %}来实现单分支选择结构。

2）双分支选择结构

使用标签{% if %}…{% else %}…{% endif %}来实现双分支选择结构。

3）多分支选择结构

使用标签{% if %}…{% elif %}…{% else %}…{% endif %}来实现多分支选择结构。

4）选择结构嵌套

对上述 3 种选择结构进行相互嵌套，即可用于表达更加复杂的选择结构。

下面通过一个示例，演示一下如何使用选择结构。

（1）创建名为 flaskProject 的 Flask 项目。

（2）打开保存在项目根目录下的文件 app.py，代码如下：

```
# 资源包\Code\chapter3\3.6\7\flaskProject\app.py
from flask import Flask, render_template
app = Flask(__name__)
# 单分支选择结构
@app.route('/single_select/')
def single_select():
    return render_template('single_select.html', username = 'xzd')
# 双分支选择结构
@app.route('/dual_select/')
def dual_select():
    return render_template('dual_select.html', username = 'oldxia')
# 双分支选择结构
@app.route('/multi_select/')
def multi_select():
    return render_template('multi_select.html', score = 95)
# 选择结构嵌套
```

```
@app.route('/nested_select/')
def nested_select():
    return render_template('nested_select.html', score = 36)
if __name__ == '__main__':
    app.run(debug = True)
```

(3) 在项目的根目录下的文件夹 templates 中创建模板文件 single_select.html, 代码如下:

```
#资源包\Code\chapter3\3.6\7\flaskProject\templates\single_select.html
<!DOCTYPE html>
<html lang = "en">
    <head>
        <meta charset = "UTF-8">
        <meta name = "viewport" content = "width = device-width, initial-scale = 1.0">
        <title>Document</title>
    </head>
    <body>
        {% if username == 'xzd' %}
            <h1>欢迎用户{{ username }}登录网站</h1>
        {% endif %}
    </body>
</html>
```

(4) 在项目的根目录下的文件夹 templates 中创建模板文件 dual_select.html, 代码如下:

```
#资源包\Code\chapter3\3.6\7\flaskProject\templates\dual_select.html
<!DOCTYPE html>
<html lang = "en">
    <head>
        <meta charset = "UTF-8">
        <meta name = "viewport" content = "width = device-width, initial-scale = 1.0">
        <title>Document</title>
    </head>
    <body>
        {% if username == 'xzd' %}
            <h1>欢迎用户{{ username }}登录网站</h1>
        {% else %}
            <h1>用户{{ username }}不是本网站用户,无法登录!</h1>
        {% endif %}
    </body>
</html>
```

(5) 在项目的根目录下的文件夹 templates 中创建模板文件 multi_select.html, 代码如下:

```
#资源包\Code\chapter3\3.6\7\flaskProject\templates\multi_select.html
<!DOCTYPE html>
<html lang = "en">
    <head>
        <meta charset = "UTF-8">
        <meta name = "viewport" content = "width = device-width, initial-scale = 1.0">
```

```
            <title>Document</title>
        </head>
        <body>
            {% if score >= 90 %}
                <h1>您的分数{{ score }}分,成绩优秀</h1>
            {% elif score >= 70 %}
                <h1>您的分数{{ score }}分,成绩良好</h1>
            {% elif score >= 60 %}
                <h1>您的分数{{ score }}分,成绩及格</h1>
            {% else %}
                <h1>您的分数{{ score }}分,成绩不及格</h1>
            {% endif %}
        </body>
</html>
```

(6) 在项目的根目录下的文件夹 templates 中创建模板文件 nested_select.html,代码如下:

```
#资源包\Code\chapter3\3.6\7\flaskProject\templates\nested_select.html
<!DOCTYPE html>
<html lang="en">
    <head>
        <meta charset="UTF-8">
        <meta name="viewport" content="width=device-width, initial-scale=1.0">
        <title>Document</title>
    </head>
    <body>
        {% if score >= 60 %}
            <h1>您的分数{{ score }}分,考试及格</h1>
        {% else %}
            {% if score >= 45 %}
                <h1>您的分数{{ score }}分,考试不及格,但可以参加补考</h1>
            {% else %}
                <h1>您的分数{{ score }}分,考试不及格,并且不可以参加补考</h1>
            {% endif %}
        {% endif %}
    </body>
</html>
```

(7) 运行上述程序,打开浏览器并在网址栏输入 http://127.0.0.1:5000/single_select/,其显示内容如图 3-49 所示。

图 3-49　浏览器中的显示内容

(8) 在浏览器中的网址栏输入 http://127.0.0.1:5000/dual_select/,其显示内容如图 3-50 所示。

图 3-50　浏览器中的显示内容

（9）在浏览器中的网址栏输入 http://127.0.0.1:5000/multi_select/，其显示内容如图 3-51 所示。

图 3-51　浏览器中的显示内容

（10）在浏览器中的网址栏输入 http://127.0.0.1:5000/nested_select/，其显示内容如图 3-52 所示。

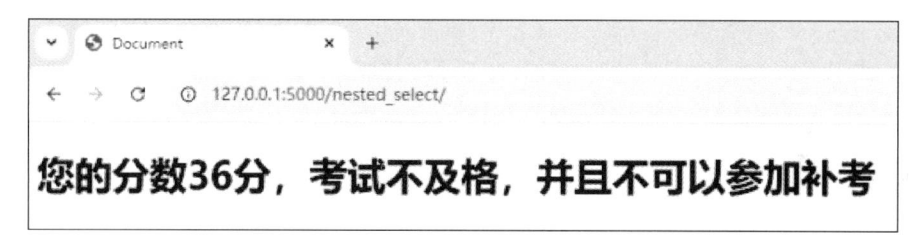

图 3-52　浏览器中的显示内容

2. 循环结构

在模板中,通过标签｛% for %｝…｛% endfor %｝或标签｛% for %｝…｛% else %｝…｛% endfor %｝来实现循环结构。

此外,在循环结构中还包含多个内置的循环变量,主要用于获取当前遍历的状态,具体如表 3-6 所示。

表 3-6　内置循环变量

内置循环变量	描述
loop.index	当前迭代的索引,从 1 开始
loop.index0	当前迭代的索引,从 0 开始
loop.first	是否是第 1 次迭代,返回值为 True 或 False
loop.last	是否是最后一次迭代,返回值为 True 或 False
loop.length	序列的长度

下面通过一个示例,演示一下如何使用循环结构。

（1）创建名为 flaskProject 的 Flask 项目。

(2) 打开保存在项目根目录下的文件 app.py,代码如下:

```python
# 资源包\Code\chapter3\3.6\8\flaskProject\app.py
from flask import Flask, render_template
app = Flask(__name__)
@app.route('/')
def index():
    context = {
        "books": [
            {
                "name": "《Python全栈开发——基础入门》",
                "author": "夏正东",
                "price": 79
            },
            {
                "name": "《Python全栈开发——高阶编程》",
                "author": "夏正东",
                "price": 89
            },
            {
                "name": "《Python全栈开发——数据分析》",
                "author": "夏正东",
                "price": 79
            }
        ]
    }
    return render_template('index.html', **context)
if __name__ == "__main__":
    app.run(debug = True)
```

(3) 在项目的根目录下的文件夹 templates 中创建模板文件 index.html,代码如下:

```html
# 资源包\Code\chapter3\3.6\8\flaskProject\templates\index.html
<!DOCTYPE html>
<html lang = "en">
    <head>
        <meta charset = "UTF-8">
        <title>Title</title>
    </head>
    <body>
        <table border = "1" cellspacing = "0">
            <tr>
                <th>序号0</th>
                <th>序号1</th>
                <th>书名</th>
                <th>作者</th>
                <th>价格</th>
            </tr>
            {% for book in books %}
                {% if loop.first %}
                    <tr style = "background: yellow">
                {% elif loop.last %}
                    <tr style = "background: greenyellow">
                {% else %}
                    <tr>
```

```
                    {% endif %}
                    <td>{{ loop.index0 }}</td>
                    <td>{{ loop.index }}</td>
                    <td>{{ book.name }}</td>
                    <td>{{ book.author }}</td>
                    <td>{{ book.price }}</td>
                </tr>
            {% endfor %}
        </table>
    </body>
</html>
```

(4) 运行上述程序,打开浏览器并在网址栏输入 http://127.0.0.1:5000/,其显示内容如图 3-53 所示。

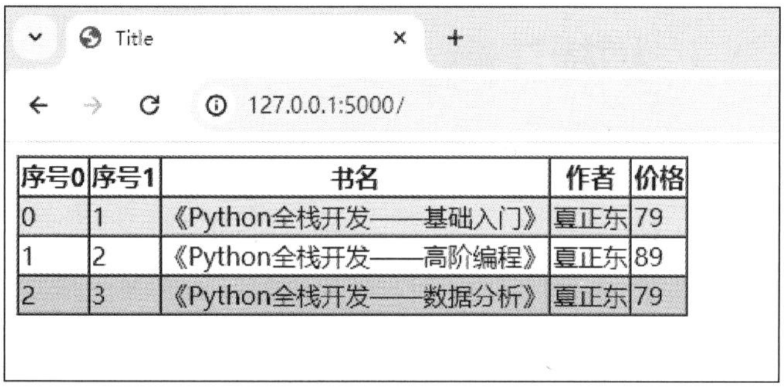

图 3-53　浏览器中的显示内容

3.6.7　模板注释

在模板中,通过标签{#…#}添加注释,需要注意的是模板注释不会出现在 HTML 文档之中。

下面通过一个示例,演示一下如何进行模板注释。

(1) 创建名为 flaskProject 的 Flask 项目。
(2) 打开保存在项目根目录下的文件 app.py,代码如下:

```
# 资源包\Code\chapter3\3.6\9\flaskProject\app.py
from flask import Flask, render_template
app = Flask(__name__)
@app.route('/')
def index():
    return render_template('index.html')
if __name__ == "__main__":
    app.run(debug = True)
```

(3) 在项目的根目录下的文件夹 templates 中创建模板文件 index.html,代码如下:

```
# 资源包\Code\chapter3\3.6\9\flaskProject\templates\index.html
<!DOCTYPE html>
<html lang = "en">
    <head>
        <meta charset = "UTF-8">
```

```html
        <title>Title</title>
    </head>
    <body>
        {# 这是模板注释 #}
        <h1>模板注释</h1>
    </body>
</html>
```

(4) 运行上述程序,打开浏览器并在网址栏输入 http://127.0.0.1:5000/,其显示内容如图 3-54 所示。

图 3-54　浏览器中的显示内容

3.6.8　宏

宏是 Jinja2 中所提供的一个非常有用的特性,类似于 Python 中的函数,可以把一部分模板代码封装到宏中,并可以通过传递的参数来构建内容。

通常情况下,可以将一些经常使用的代码片段放到宏中,并将不固定的值抽取出来当作变量,进而达到减少代码量,避免代码冗余的目的。

1. 创建宏

在模板中,通过标签{% macro %}…{% endmacro %}进行宏的创建。

下面通过一个示例,演示一下如何创建宏。

(1) 创建名为 flaskProject 的 Flask 项目。

(2) 打开保存在项目根目录下的文件 app.py,代码如下:

```python
# 资源包\Code\chapter3\3.6\10\flaskProject\app.py
from flask import Flask, render_template
app = Flask(__name__)
@app.route('/')
def index():
    return render_template('index.html')
if __name__ == "__main__":
    app.run(debug = True)
```

(3) 在项目的根目录下的文件夹 templates 中创建模板文件 index.html,代码如下:

```html
# 资源包\Code\chapter3\3.6\10\flaskProject\templates\index.html
{# 创建宏 #}
{% macro input(tip,name,type) %}
    {{ tip }}<input name="{{ name }}" type="{{ type }}">
{% endmacro %}
<!DOCTYPE html>
```

```html
<html lang = "en">
    <head>
        <meta charset = "UTF-8">
        <title>老夏学院</title>
    </head>
    <body>
        <table>
            <tr>
                {# 应用宏 #}
                <td>{{ input('账号: ','username','text') }}</td>
            </tr>
            <tr>
                {# 应用宏 #}
                <td>{{ input('密码: ','password','password') }}</td>
            </tr>
            <tr>
                {# 应用宏 #}
                <td>{{ input('','提交','submit') }}</td>
            </tr>
        </table>
    </body>
</html>
```

（4）运行上述程序，打开浏览器并在网址栏输入 http://127.0.0.1:5000/，其显示内容如图 3-55 所示。

2. 导入宏

为了便于管理，通常将宏存储在单独的文件中，并将该文件命名为 macros.html 或 _macros.html。

图 3-55　浏览器中的显示内容

而在模板中，则可以通过标签{% import…as…%}、标签{% from…import… %}或标签{% from…import…as…%}将指定的宏导入。

这里需要注意的是，导入宏文件的不要使用相对路径，而应使用以文件夹 templates 为根目录的路径。

下面通过一个示例，演示一下如何导入宏。

（1）创建名为 flaskProject 的 Flask 项目。

（2）打开保存在项目根目录下的文件 app.py，代码如下：

```
# 资源包\Code\chapter3\3.6\11\flaskProject\app.py
from flask import Flask, render_template
app = Flask(__name__)
@app.route('/')
def index():
    return render_template('index.html')
if __name__ == "__main__":
    app.run(debug = True)
```

（3）在项目的根目录下的文件夹 templates 中创建模板文件 index.html，代码如下：

```
#资源包\Code\chapter3\3.6\11\flaskProject\templates\index.html
{% from "macros/macros.html" import input as input_field %}
{#{% import "macros/macros.html" as macros %}#}
<!DOCTYPE html>
<html lang="en">
    <head>
        <meta charset="UTF-8">
        <title>老夏学院</title>
    </head>
    <body>
        <table>
            <tr>
                <td>{{ input_field('账号：','username','text') }}</td>
                {#<td>{{ macros.input('账号：','username','text') }}</td>#}
            </tr>
            <tr>
                <td>{{ input_field('密码：','password','password') }}</td>
                {#<td>{{ macros.input('密码：','password','password') }}</td>#}
            </tr>
            <tr>
                <td>{{ input_field('','提交','submit') }}</td>
                {#<td>{{ macros.input('','提交','submit') }}</td>#}
            </tr>
        </table>
    </body>
</html>
```

（4）在项目的根目录下的文件夹 templates 中创建文件夹 macros，并在其中创建宏文件 macros.html，代码如下：

```
#资源包\Code\chapter3\3.6\11\flaskProject\templates\macros\macros.html
{% macro input(tip,name,type) %}
    {{ tip }}<input name="{{ name }}" type="{{ type }}">
{% endmacro %}
```

（5）运行上述程序，打开浏览器并在网址栏输入 http://127.0.0.1:5000/，其显示内容如图 3-56 所示。

图 3-56　浏览器中的显示内容

3.6.9　include 标签

该标签用于将一个模板引入另外一个模板中的指定位置，通过标签{% include… %}实现。

此外,通过 include 标签引入模板的路径与导入宏一样,不要使用相对路径,而应以文件夹 templates 为根目录。

下面通过一个示例,演示一下如何使用 include 标签。

(1) 创建名为 flaskProject 的 Flask 项目。

(2) 打开保存在项目根目录下的文件 app.py,代码如下:

```
# 资源包\Code\chapter3\3.6\12\flaskProject\app.py
from flask import Flask, render_template
app = Flask(__name__)
@app.route('/')
def index():
    return render_template('index.html')
if __name__ == "__main__":
    app.run(debug = True)
```

(3) 在项目的根目录下的文件夹 templates 中创建模板文件 index.html,代码如下:

```
# 资源包\Code\chapter3\3.6\12\flaskProject\templates\index.html
{% from "macros/macros.html" import input as input_field %}
{#{% import "macros/macros.html" as macros %}#}
<!DOCTYPE html>
<html lang = "en">
    <head>
        <meta charset = "UTF-8">
        <title>老夏学院</title>
    </head>
    <body>
        {% include "common/header.html" %}
        <table>
            <tr>
                <td>{{ input_field('账号:','username','text') }}</td>
                {#<td>{{ macros.input('账号:','username','text') }}</td>#}
            </tr>
            <tr>
                <td>{{ input_field('密码:','password','password') }}</td>
                {#<td>{{ macros.input('密码:','password','password') }}</td>#}
            </tr>
            <tr>
                <td>{{ input_field('','提交','submit') }}</td>
                {#<td>{{ macros.input('','提交','submit') }}</td>#}
            </tr>
        </table>
        {% include "common/footer.html" %}
    </body>
</html>
```

(4) 在项目的根目录下的文件夹 templates 中创建文件夹 macros,并在其中创建宏文件 macros.html,代码如下:

```
# 资源包\Code\chapter3\3.6\12\flaskProject\templates\macros\macros.html
{% macro input(tip,name,type) %}
    {{ tip }}<input name = "{{ name }}" type = "{{ type }}">
{% endmacro %}
```

（5）在项目的根目录下的文件夹 templates 中创建文件夹 common，并在其中创建模板文件 header.html，代码如下：

```
# 资源包\Code\chapter3\3.6\12\flaskProject\templates\common\header.html
< h1 style = "background: red">这是 header </h1 >
```

（6）在项目的根目录下的文件夹 templates 中的文件夹 common 中创建模板文件 footer.html，代码如下：

```
# 资源包\Code\chapter3\3.6\12\flaskProject\templates\common\footer.html
< h1 style = "background: green">这是 footer </h1 >
```

（7）运行上述程序，打开浏览器并在网址栏输入 http://127.0.0.1:5000/，其显示内容如图 3-57 所示。

图 3-57　浏览器中的显示内容

3.6.10　set 语句和 with 语句

在模板中，set 语句和 with 语句主要用于定义模板变量。
1）set 语句
通过标签{% set… %}定义模板变量，其作用域为整段代码。
2）with 语句
通过标签{% with… %}…{% endwith %}定义模板变量，其作用域为 with 语句内。
这里需要注意的是，当 set 语句嵌入 with 语句中使用时，其作用域则从整段代码变为仅在 with 语句内。
下面通过一个示例，演示一下如何使用 set 语句和 with 语句。
（1）创建名为 flaskProject 的 Flask 项目。
（2）打开保存在项目根目录下的文件 app.py，代码如下：

```
# 资源包\Code\chapter3\3.6\13\flaskProject\app.py
from flask import Flask, render_template
app = Flask(__name__)
@app.route('/')
def index():
```

```
        return render_template('index.html')
if __name__ == "__main__":
    app.run(debug = True)
```

（3）在项目的根目录下的文件夹 templates 中创建模板文件 index.html，代码如下：

```
# 资源包\Code\chapter3\3.6\13\flaskProject\templates\index.html
<!DOCTYPE html>
<html lang = "en">
    <head>
        <meta charset = "UTF-8">
        <title>老夏学院</title>
    </head>
    <body>
        {% set author = "夏正东" %}
        <p>《Python全栈开发——Web编程》的作者：{{ author }}</p>
        {% with website = "http://www.oldxia.com" %}
            <p>老夏学院的网址：{{ website }}</p>
            {% set author_tel = "13309XXXXXX" %}
            <p style = "background: antiquewhite">作者的联系方式：{{ author_tel }}</p>
        {% endwith %}
        <p style = "background: yellow">作者的联系方式：{{ author_tel }}</p>
    </body>
</html>
```

（4）运行上述程序，打开浏览器并在网址栏输入 http://127.0.0.1:5000/，其显示内容如图 3-58 所示。

图 3-58　浏览器中的显示内容

3.6.11　加载静态文件

常用的静态文件主要有 CSS 文件、JavaScript 脚本和图片等。在模板中，静态文件默认存放在项目的根目录下的 static 文件夹中，可以通过 url_for() 函数加载。

此外，可以在创建 Flask 实例对象时，通过添加参数 static_folder 来指定静态文件存放的位置。

下面通过一个示例，演示一下如何加载静态文件。

（1）创建名为 flaskProject 的 Flask 项目。

（2）打开保存在项目根目录下的文件 app.py，代码如下：

```
# 资源包\Code\chapter3\3.6\14\flaskProject\app.py
from flask import Flask, render_template
app = Flask(__name__)
@app.route('/')
def index():
    return render_template('index.html')
if __name__ == "__main__":
    app.run(debug = True)
```

(3) 在项目的根目录下的文件夹 templates 中创建模板文件 index.html,代码如下:

```
# 资源包\Code\chapter3\3.6\14\flaskProject\templates\index.html
<!DOCTYPE html>
<html lang="en">
    <head>
        <meta charset="UTF-8">
        <title>老夏学院</title>
        <link rel="stylesheet" href="{{ url_for('static', filename = 'css/index.css') }}">
        <script src="{{url_for('static', filename = 'js/index.js')}}"></script>
    </head>
    <body>
        <h1>老夏学院:www.oldxia.com</h1>
        <img src="{{ url_for('static', filename = 'images/oldxia.png') }}">
    </body>
</html>
```

(4) 在项目的根目录下的文件夹 static 中创建文件夹 css,并在其中创建文件 index.css,代码如下:

```
# 资源包\Code\chapter3\3.6\14\flaskProject\static\css\index.css
h1{
    font-size: 30px;
    font-weight: bold;
    color: red;
}
```

(5) 在项目的根目录下的文件夹 static 中创建文件夹 js,并在其中创建文件 index.js,代码如下:

```
# 资源包\Code\chapter3\3.6\14\flaskProject\static\js\index.js
alert("感谢您购买《Python全栈开发——Web编程》")
```

(6) 在项目的根目录下的文件夹 static 中创建文件夹 images,并将图片数据存放其中。

(7) 运行上述程序,打开浏览器并在网址栏输入 http://127.0.0.1:5000/,其显示内容如图 3-59 所示。

图 3-59　浏览器中的显示内容

3.6.12 模板继承

模板继承是Jinja2的重要特性之一。通过模板继承,可以将模板中重复出现的元素提取出来,并存放在一个已定义的父模板之中,进而达到避免重复编写代码的目的。

模板继承可以分为3步。

(1) 在父模板中,通过标签{% block…%}…{% endblock %}保存Web页面中的常用元素。

(2) 在子模板中使用标签{% extends…%}继承父模板,这里需要注意的是,父模板的路径要以文件夹templates为根目录。

(3) 在子模板中使用标签{% block…%}…{% endblock %}将子模板中的内容插入父模板中,需要注意的是,子模板标签{% block… %}…{% endblock %}中的内容会覆盖父模板中的内容。

此外还需要注意3点:一是如果需要保留父模板标签{% block… %}…{% endblock %}中的内容,则需要在子模板标签{% block… %}…{% endblock %}之中使用标签{{ super() }};二是如果需要使用子模板标签{% block… %}…{% endblock %}中的内容,则需要使用标签{{ self.block 名称() }}进行调用;三是子模板标签{% block…%}…{% endblock %}之外的代码,将不会被模板引擎渲染。

下面通过一个示例,演示一下如何使用模板继承。

(1) 创建名为flaskProject的Flask项目。

(2) 打开保存在项目根目录下的文件app.py,代码如下:

```
# 资源包\Code\chapter3\3.6\15\flaskProject\app.py
from flask import Flask, render_template
app = Flask(__name__)
@app.route('/')
def index():
    return render_template('index.html')
@app.route('/detail/')
def detail():
    return render_template('detail.html')
if __name__ == "__main__":
    app.run(debug = True)
```

(3) 在项目的根目录下的文件夹templates中创建父模板文件base.html,代码如下:

```
# 资源包\Code\chapter3\3.6\15\flaskProject\templates\base.html
<!DOCTYPE html>
<html lang = "en">
    <head>
        <meta charset = "UTF-8">
        <title>{% block title %}老夏学院{% endblock %}</title>
    </head>
    <body>
        {% block head %}
            <h3>网站顶部内容(父模板)</h3>
        {% endblock %}
        {% block main %}
```

```
            <h1>网站主体部分(父模板)</h1>
        {% endblock %}
        {% block footer %}
            <h3>网站底部内容(父模板)</h3>
        {% endblock %}
    </body>
</html>
```

（4）在项目的根目录下的文件夹 templates 中创建子模板文件 index.html，代码如下：

```
#资源包\Code\chapter3\3.6\15\flaskProject\templates\index.html
{% extends "base.html" %}
{% block main %}
    {{ super() }}
    <h1 style="background: greenyellow">网站主题内容(子模板 index)</h1>
{% endblock %}
```

（5）在项目的根目录下的文件夹 templates 中创建子模板文件 detail.html，代码如下：

```
#资源包\Code\chapter3\3.6\15\flaskProject\templates\detail.html
{% extends "base.html" %}
{% block title %}
    老夏学院-详情页
{% endblock %}
{% block main %}
    {{ self.title() }}
    <h1 style="background: aqua">网站主体部分(子模板 detail)</h1>
{% endblock %}
老夏学院：http://www.oldxia.com
```

（6）运行上述程序，打开浏览器并在网址栏输入 http://127.0.0.1:5000/，其显示内容如图 3-60 所示。

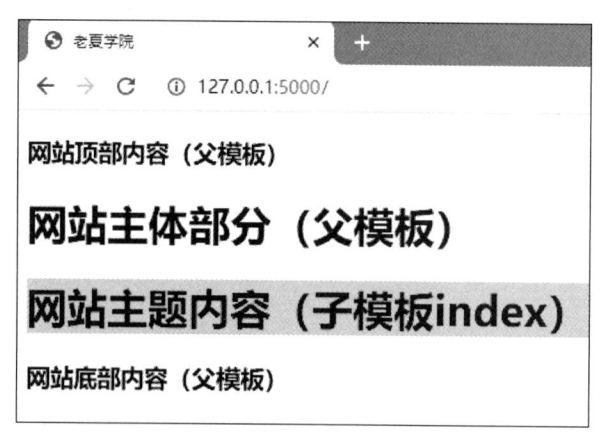

图 3-60　浏览器中的显示内容

（7）在浏览器中的网址栏输入 http://127.0.0.1:5000/detail/，其显示内容如图 3-61 所示。

图 3-61 浏览器中的显示内容

3.7 类视图

由于前面章节所使用的视图都是函数,所以称为视图函数,其实,视图也可以基于类来实现,而类视图的好处就是支持继承,即可将共性的内容抽取出来放到父类中,然后在子类中完成各自的业务逻辑,并继承父类。

此外,需要注意的是,类视图不同于视图函数,其需要使用 add_url_rule() 方法进行注册。

类视图分为标准类视图和基于调度方法的类视图。

1. 标准类视图

标准类视图继承自 flask.views 模块中的 View 类,在标准类视图中必须重写 dispatch_request() 方法,这种方法类似于视图函数,将会返回一个基于 Response 类或者其子类的对象。

标准类视图创建后,需要通过 add_url_rule() 方法与路由进行映射,需要注意的是,该方法中的参数 view_func 不能直接传入类的名称,而是需要使用 View 类的类方法 as_view() 将类转换成可以为路由注册的视图函数,并在其中指定 URL 的名称,以便于 url_for() 函数的调用。

下面通过一个示例,演示一下如何使用标准类视图。

(1) 创建名为 flaskProject 的 Flask 项目。

(2) 打开保存在项目根目录下的文件 app.py,代码如下:

```
# 资源包\Code\chapter3\3.7\1\flaskProject\app.py
from flask import Flask, views, render_template
app = Flask(__name__)
class Index(views.View):
    def dispatch_request(self):
        return render_template('index.html')
app.add_url_rule('/', view_func = Index.as_view('index'))
class UserView(views.View):
    def __init__(self):
```

```python
        super().__init__()
        self.context = {
            'sitename': '老夏学院'
        }
class LoginView(UserView):
    def dispatch_request(self):
        self.context.update({
            'siteurl': 'www.oldxia.com'
        })
        return render_template('login.html', **self.context)
class RegistView(UserView):
    def dispatch_request(self):
        return render_template('regist.html', **self.context)
app.add_url_rule('/login/', endpoint = 'my_login', view_func = LoginView.as_view('login'))
app.add_url_rule('/regist/', view_func = RegistView.as_view('regist'))
if __name__ == "__main__":
    app.run(debug = True)
```

（3）在项目的根目录下的文件夹 templates 中创建模板文件 index.html，代码如下：

```
#资源包\Code\chapter3\3.7\1\flaskProject\templates\index.html
<!DOCTYPE html>
<html lang = "en">
    <head>
        <meta charset = "UTF-8">
        <title>老夏学院</title>
    </head>
    <body>
        <h1>老夏学院的主页</h1>
        <a href = "{{ url_for('my_login') }}">登录网站</a><br>
        <a href = "{{ url_for('regist') }}">注册账号</a>
    </body>
</html>
```

（4）在项目的根目录下的文件夹 templates 中创建模板文件 login.html，代码如下：

```
#资源包\Code\chapter3\3.7\1\flaskProject\templates\login.html
<!DOCTYPE html>
<html lang = "en">
    <head>
        <meta charset = "UTF-8">
        <title>Document</title>
    </head>
    <body>
        <h1>登录页面</h1>
        <h1 style = "color:red">{{ sitename }}</h1>
        <h1 style = "color:red">{{ siteurl }}</h1>
    </body>
</html>
```

（5）在项目的根目录下的文件夹 templates 中创建模板文件 regist.html，代码如下：

```
#资源包\Code\chapter3\3.7\1\flaskProject\templates\regist.html
<!DOCTYPE html>
<html lang = "en">
```

```html
    <head>
        <meta charset = "UTF-8">
        <title>Title</title>
    </head>
    <body>
        <h1>注册页面</h1>
        <h1 style = "color:red">{{ sitename }}</h1>
    </body>
</html>
```

（6）运行上述程序，打开浏览器并在网址栏输入 http://127.0.0.1:5000/，其显示内容如图 3-62 所示。

（7）单击"登录网站"按钮，其显示内容如图 3-63 所示。

图 3-62　浏览器中的显示内容 　　　　图 3-63　浏览器中的显示内容

（8）单击"注册账号"按钮，其显示内容如图 3-64 所示。

2．基于调度方法的类视图

基于调度方法的类视图继承自 flask.view 模块中的 MethodView 类，其根据不同的 HTTP 请求来执行不同的方法，例如：当用户发送 GET 请求时，将会执行该类中的 get()方法；当用户发送 POST 请求时，将会执行该类中的 post()方法。

图 3-64　浏览器中的显示内容

下面通过一个示例，演示一下如何使用基于调度方法的类视图。

（1）创建名为 flaskProject 的 Flask 项目。

（2）打开保存在项目根目录下的文件 app.py，代码如下：

```python
#资源包\Code\chapter3\3.7\2\flaskProject\app.py
from flask import Flask, views, render_template, request
app = Flask(__name__)
class LoginView(views.MethodView):
    def __render(self, errorinfo = None):
        return render_template('login.html', errorinfo = errorinfo)
    def get(self):
        return self.__render()
    def post(self):
```

```python
        username = request.form.get('username')
        password = request.form.get('password')
        if username == 'oldxia' and password == '123456':
            return render_template('index.html', username = username)
        else:
            return self.__render(errorinfo = '用户名或密码错误,请确认后重新登录')
app.add_url_rule('/login/', endpoint = 'my_login', view_func = LoginView.as_view('login'))
if __name__ == "__main__":
    app.run(debug = True)
```

(3) 在项目的根目录下的文件夹 templates 中创建模板文件 login.html,代码如下:

```html
# 资源包\Code\chapter3\3.7\2\flaskProject\templates\login.html
<!DOCTYPE html>
<html lang = "en">
    <head>
        <meta charset = "UTF-8">
        <title>Document</title>
    </head>
    <body>
        <form action = "" method = "post">
            <table>
                <tr>
                    <td>用户名:</td>
                    <td><input type = "text" name = "username"></td>
                </tr>
                <tr>
                    <td>密码:</td>
                    <td><input type = "text" name = "password"></td>
                </tr>
                <tr>
                    <td colspan = " = 2"><input type = "submit" value = "登录"></td>
                </tr>
            </table>
            {% if errorinfo %}
                <p style = "color: red">{{ errorinfo }}</p>
            {% endif %}
        </form>
    </body>
</html>
```

(4) 在项目的根目录下的文件夹 templates 中创建模板文件 index.html,代码如下:

```html
# 资源包\Code\chapter3\3.7\2\flaskProject\templates\index.html
<!DOCTYPE html>
<html lang = "en">
    <head>
        <meta charset = "UTF-8">
        <title>老夏学院</title>
    </head>
    <body>
        <h1>欢迎用户{{ username }},登录老夏学院</h1>
    </body>
</html>
```

(5) 运行上述程序，打开浏览器并在网址栏输入 http://127.0.0.1:5000/login/，其显示内容如图 3-65 所示。

此时，当输入正确的用户名和密码时，其登录后的显示内容如图 3-66 所示。

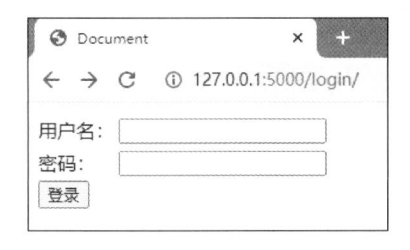

图 3-65　浏览器中的显示内容

图 3-66　浏览器中的显示内容

而当输入错误的用户名或密码时，其登录后的显示内容如图 3-67 所示。

除此之外，在类视图中，可以通过重写类视图中的类属性 decorators 来自定义装饰器。

下面通过一个示例，演示一下如何自定义类视图装饰器。

(1) 创建名为 flaskProject 的 Flask 项目。

(2) 打开保存在项目根目录下的文件 app.py，代码如下：

图 3-67　浏览器中的显示内容

```python
#资源包\Code\chapter3\3.7\3\flaskProject\app.py
import functools
from flask import Flask, request, render_template, views
app = Flask(__name__)
#自定义的函数装饰器
def login_required(func):
    @functools.wraps(func)
    def wraps(*args, **kwargs):
        #func 表示被自定义函数装饰器所装饰的视图函数
        print(func)
        username = request.args.get('username')
        if username and username == 'oldxia':
            return func(*args, **kwargs)
        else:
            return render_template('login.html', errorinfo='先登录!')
    return wraps
class CourseView(views.View):
    decorators = [login_required]
    def dispatch_request(self):
        return "欢迎进入选课中心"
class SettingsView(views.View):
    decorators = [login_required]
    def dispatch_request(self):
        return "欢迎进入个人中心"
app.add_url_rule(rule='/course/', view_func=CourseView.as_view('course'))
app.add_url_rule(rule='/settings/', view_func=SettingsView.as_view('settings'))
if __name__ == '__main__':
    app.run(debug=True)
```

（3）在项目的根目录下的文件夹 templates 中创建模板文件 login.html，代码如下：

```
#资源包\Code\chapter3\3.7\3\flaskProject\templates\login.html
<!DOCTYPE html>
<html lang="en">
    <head>
        <meta charset="UTF-8">
        <title>Document</title>
    </head>
    <body>
        <h1 style="color: red">{{ errorinfo }}</h1>
    </body>
</html>
```

（4）运行上述程序，打开浏览器并在网址栏输入 http://127.0.0.1:5000/settings/，其显示内容如图 3-68 所示。

（5）在浏览器中的网址栏输入 http://127.0.0.1:5000/settings/?username=oldxia，模拟登录成功时的场景，其显示内容如图 3-69 所示。

图 3-68　浏览器中的显示内容　　　　图 3-69　浏览器中的显示内容

3.8　蓝图

在实际项目开发过程中，需要实现的功能非常多，这就导致业务视图也会非常多，而如果将所有的业务视图都写在同一个文件中，则在功能上是没有问题的，但却非常不便于代码的管理和后期功能代码的添加。

例如，当开发一个购物网站时，从业务角度上，可以将整个应用划分为用户模块单元、商品模块单元和订单模块单元，那么，开发人员如何做到分别开发这些不同的单元，并最终整合到一个项目的应用之中呢？

此时，就可以通过 Flask 中的蓝图进行模块化管理，进而使项目的结构更加简洁、清晰。

蓝图是一种可重用的应用程序组件，可以帮助开发人员组织和管理应用程序的路由和视图。可以将应用程序拆分成多个模块，每个模块都可以拥有自己的路由和视图，进而使应用程序更易于维护和扩展。

3.8.1　应用蓝图

在 Flask 中，蓝图的应用可以分为 3 个步骤。

（1）通过 Flask 模块中的 Blueprint 类创建蓝图，并获得蓝图对象，其语法格式如下：

```
Blueprint(name, import_name, url_prefix)
```

其中，参数 name 表示蓝图的名称，参数 import_name 表示蓝图所在的模块，参数 url_prefix 表示蓝图的 URL 前缀。

（2）在蓝图中进行路由声明等相关操作。

（3）通过 Flask 对象的 register_blueprint()方法在应用程序中注册蓝图，其语法格式如下：

```
register_blueprint(blueprint)
```

其中，参数 blueprint 表示需要被注册蓝图的名称。

下面通过一个示例，演示一下如何应用蓝图。

（1）创建名为 flaskProject 的 Flask 项目。

（2）打开保存在项目根目录下的文件 app.py，代码如下：

```
# 资源包\Code\chapter3\3.8\1\flaskProject\app.py
from flask import Flask
from blueprint.user import user
from blueprint.goods import goods
app = Flask(__name__)
# 用户模块
app.register_blueprint(blueprint = user)
# 商品模块
app.register_blueprint(blueprint = goods)
if __name__ == '__main__':
    app.run(debug = True)
```

（3）在项目的根目录下创建文件夹 blueprint，并在其中创建蓝图文件 user.py，代码如下：

```
# 资源包\Code\chapter3\3.8\1\flaskProject\blueprint\user.py
from flask import Blueprint
user = Blueprint('user', __name__)
@user.route('/profile/')
def profile():
    return '个人中心页面'
@user.route('/settings/')
def settings():
    return '个人设置页面'
```

（4）在项目的根目录下的文件夹 blueprint 中创建蓝图文件 goods.py，代码如下：

```
# 资源包\Code\chapter3\3.8\1\flaskProject\blueprint\goods.py
from flask import Blueprint
goods = Blueprint('goods', __name__, url_prefix = '/book')
@goods.route('/python/')
def python():
    return '商品——Python 类书籍'
```

（5）运行上述程序，打开浏览器并在网址栏输入 http://127.0.0.1:5000/profile/，其显示内容如图 3-70 所示。

（6）在浏览器中的网址栏输入 http://127.0.0.1:5000/settings/，其显示内容如

图 3-71 所示。

图 3-70　浏览器中的显示内容 　　　　图 3-71　浏览器中的显示内容

（7）在浏览器中的网址栏输入 http://127.0.0.1:5000/book/python/，其显示内容如图 3-72 所示。

图 3-72　浏览器中的显示内容

3.8.2　蓝图中加载模板

在蓝图中加载模板有两种方式，一是在项目中的文件夹 templates 中编写相对应的模板文件；二是在创建蓝图时为 Blueprint 类添加 template_folder 属性，用于指定模板的位置。

这里需要注意的是，如果文件夹 templates 中的模板文件与 Blueprint 类中 template_folder 属性所指定的模板路径相同，则以文件夹 templates 中的模板为主。

下面通过一个示例，演示一下如何在蓝图中加载模板。

（1）创建名为 flaskProject 的 Flask 项目。

（2）打开保存在项目根目录下的文件 app.py，代码如下：

```
# 资源包\Code\chapter3\3.8\2\flaskProject\app.py
from flask import Flask
from blueprint.user import user
from blueprint.goods import goods
from blueprint.order import order
app = Flask(__name__)
# 用户模块
app.register_blueprint(blueprint = user)
# 商品模块
app.register_blueprint(blueprint = goods)
# 订单模块
app.register_blueprint(blueprint = order)
if __name__ == '__main__':
    app.run(debug = True)
```

（3）在项目的根目录下创建文件夹 blueprint，并在其中创建蓝图文件 user.py，代码如下：

```
# 资源包\Code\chapter3\3.8\2\flaskProject\blueprint\user.py
from flask import render_template
```

```python
from flask import Blueprint
user = Blueprint('user', __name__)
@user.route('/profile/')
def profile():
    return '个人中心页面'
@user.route('/settings/')
def settings():
    return render_template('settings.html')
```

(4) 在项目的根目录下的文件夹 blueprint 中创建蓝图文件 goods.py，代码如下：

```python
# 资源包\Code\chapter3\3.8\2\flaskProject\blueprint\goods.py
from flask import render_template
from flask import Blueprint
goods = Blueprint('goods', __name__, url_prefix='/book', template_folder='tep')
@goods.route('/python/')
def python():
    return render_template('book.html')
```

(5) 在项目的根目录下的文件夹 blueprint 中创建蓝图文件 order.py，代码如下：

```python
# 资源包\Code\chapter3\3.8\2\flaskProject\blueprint\order.py
from flask import render_template
from flask import Blueprint
order = Blueprint('order', __name__, template_folder='tep')
@order.route('/sell/')
def sell():
    return render_template('sell.html')
```

(6) 在项目的根目录下的文件夹 templates 中创建模板文件 settings.html，代码如下：

```html
# 资源包\Code\chapter3\3.8\2\flaskProject\templates\settings.html
<!DOCTYPE html>
<html lang="en">
    <head>
        <meta charset="UTF-8">
    </head>
    <body>
        个人设置页面
    </body>
</html>
```

(7) 在项目的根目录下的文件夹 templates 中创建模板文件 book.html，代码如下：

```html
# 资源包\Code\chapter3\3.8\2\flaskProject\templates\book.html
<!DOCTYPE html>
<html lang="en">
    <head>
        <meta charset="UTF-8">
    </head>
    <body>
        商品——Python类书籍(templates中的模板文件)
    </body>
</html>
```

(8) 在项目的根目录下的文件夹 blueprint 中创建文件夹 tep，并在其中创建模板文件 book.html，代码如下：

```html
# 资源包\Code\chapter3\3.8\2\flaskProject\blueprint\tep\book.html
<!DOCTYPE html>
<html lang = "en">
    <head>
        <meta charset = "UTF-8">
    </head>
    <body>
        商品——Python类书籍(template_folder 属性所指定的路径)
    </body>
</html>
```

(9) 在项目的根目录下的文件夹 blueprint 中的文件夹 tep 中创建模板文件 sell.html，代码如下：

```html
# 资源包\Code\chapter3\3.8\2\flaskProject\blueprint\tep\sell.html
<!DOCTYPE html>
<html lang = "en">
    <head>
        <meta charset = "UTF-8">
    </head>
    <body>
        销售订单页面
    </body>
</html>
```

(10) 运行上述程序，打开浏览器并在网址栏输入 http://127.0.0.1:5000/settings/，其显示内容如图 3-73 所示。

(11) 在浏览器中的网址栏输入 http://127.0.0.1:5000/book/python/，其显示内容如图 3-74 所示。

图 3-73　浏览器中的显示内容

图 3-74　浏览器中的显示内容

(12) 在浏览器中的网址栏输入 http://127.0.0.1:5000/sell/，其显示内容如图 3-75 所示。

图 3-75　浏览器中的显示内容

3.8.3　蓝图中加载静态文件

在蓝图中加载静态文件的方式有两种，一是从在项目根目录下的文件夹 static 中加载相对应的静态文件；二是在创建蓝图时为 Blueprint 类添加 static_folder 属性，用于指定静态文件的位置，需要注意的是，在加载静态文件时，url_for()函数中的第 1 个参数值需为"蓝图名称.静态文件所在文件夹"的形式，这样才可以加载 static_folder 属性所指定文件夹下的静态文件。

下面通过一个示例，演示一下如何在蓝图中加载静态文件。

（1）创建名为 flaskProject 的 Flask 项目。

（2）打开保存在项目根目录下的文件 app.py，代码如下：

```
# 资源包\Code\chapter3\3.8\3\flaskProject\app.py
from flask import Flask
from blueprint.user import user
from blueprint.goods import goods
app = Flask(__name__)
# 用户模块
app.register_blueprint(blueprint = user)
# 商品模块
app.register_blueprint(blueprint = goods)
if __name__ == '__main__':
    app.run(debug = True)
```

（3）在项目的根目录下创建文件夹 blueprint，并在其中创建蓝图文件 user.py，代码如下：

```
# 资源包\Code\chapter3\3.8\3\flaskProject\blueprint\user.py
from flask import render_template
from flask import Blueprint
user = Blueprint('user', __name__)
@user.route('/settings/')
def settings():
    return render_template('settings.html')
```

（4）在项目的根目录下的文件夹 blueprint 中创建蓝图文件 goods.py，代码如下：

```
# 资源包\Code\chapter3\3.8\3\flaskProject\blueprint\goods.py
from flask import render_template
from flask import Blueprint
goods = Blueprint('goods', __name__, url_prefix = '/book', static_folder = 'sta')
@goods.route('/python/')
def python():
    return render_template('book.html')
```

（5）在项目的根目录下的文件夹 templates 中创建模板文件 settings.html，代码如下：

```
# 资源包\Code\chapter3\3.8\3\flaskProject\templates\settings.html
<!DOCTYPE html>
< html lang = "en">
    < head >
```

```html
        <meta charset="UTF-8">
        <link rel="stylesheet" href="{{ url_for('static',filename='index.css')}}">
    </head>
    <body>
        个人设置页面
    </body>
</html>
```

(6) 在项目的根目录下的文件夹 templates 中创建模板文件 book.html, 代码如下:

```html
#资源包\Code\chapter3\3.8\3\flaskProject\templates\book.html
<!DOCTYPE html>
<html lang="en">
    <head>
        <meta charset="UTF-8">
        <link rel="stylesheet" href="{{ url_for('goods.static',filename='index.css')}}">
    </head>
    <body>
        商品——Python 类书籍(templates 中的模板文件)
    </body>
</html>
```

(7) 在项目的根目录下的文件夹 static 中创建文件 index.css, 代码如下:

```css
#资源包\Code\chapter3\3.8\3\flaskProject\static\index.css
body {
    background: pink;
}
```

(8) 在项目的根目录下的文件夹 blueprint 中创建文件夹 sta, 并在其中创建文件 index.css, 代码如下:

```css
#资源包\Code\chapter3\3.8\3\flaskProject\blueprint\sta\index.css
body {
    background: yellow;
}
```

(9) 运行上述程序, 打开浏览器并在网址栏输入 http://127.0.0.1:5000/settings/, 其显示内容如图 3-76 所示。

(10) 在浏览器中的网址栏输入 http://127.0.0.1:5000/book/python/, 其显示内容如图 3-77 所示。

图 3-76　浏览器中的显示内容

图 3-77　浏览器中的显示内容

3.8.4　在蓝图中动态构建请求 URL

在蓝图中使用 url_for() 函数动态地构建请求 URL 时, 必须在视图函数之前加上蓝图的名称，即使是在同一个蓝图中，也必须指定蓝图的名称。

下面通过一个示例，演示一下如何在蓝图中动态地构建请求 URL。

(1) 创建名为 flaskProject 的 Flask 项目。

(2) 打开保存在项目根目录下的文件 app.py，代码如下：

```python
# 资源包\Code\chapter3\3.8\4\flaskProject\app.py
from flask import Flask, url_for, render_template
from blueprint.user import user
from blueprint.order import order
app = Flask(__name__)
# 用户模块
app.register_blueprint(blueprint = user)
# 订单模块
app.register_blueprint(blueprint = order)
@app.route('/')
def index():
    return render_template('index.html', user_url = url_for('user.settings'))
if __name__ == '__main__':
    app.run(debug = True)
```

(3) 在项目的根目录下创建文件夹 blueprint，并在其中创建蓝图文件 user.py，代码如下：

```python
# 资源包\Code\chapter3\3.8\4\flaskProject\blueprint\user.py
from flask import render_template
from flask import Blueprint
user = Blueprint('user', __name__)
@user.route('/settings/')
def settings():
    return render_template('settings.html')
```

(4) 在项目的根目录下的文件夹 blueprint 中创建蓝图文件 order.py，代码如下：

```python
# 资源包\Code\chapter3\3.8\4\flaskProject\blueprint\order.py
from flask import render_template, url_for
from flask import Blueprint
order = Blueprint('order', __name__, template_folder = 'tep')
@order.route('/sell/')
def sell():
    return render_template('sell.html', order_url = url_for('order.sell'))
```

(5) 在项目的根目录下的文件夹 templates 中创建模板文件 index.html，代码如下：

```html
# 资源包\Code\chapter3\3.8\4\flaskProject\templates\index.html
<!DOCTYPE html>
<html lang = "en">
    <head>
        <meta charset = "UTF-8">
    </head>
    <body>
        <p>个人设置页面的 URL：{{ user_url }}</p>
        <a href = "{{ url_for('user.settings') }}">个人设置页面</a><br>
        <a href = "{{ url_for('order.sell') }}">销售订单</a>
    </body>
</html>
```

(6) 在项目的根目录下的文件夹 templates 中创建模板文件 settings.html,代码如下:

```html
# 资源包\Code\chapter3\3.8\4\flaskProject\templates\settings.html
<!DOCTYPE html>
<html lang = "en">
    <head>
        <meta charset = "UTF-8">
    </head>
    <body>
        个人设置页面
    </body>
</html>
```

(7) 在项目的根目录下的文件夹 templates 中创建模板文件 sell.html,代码如下:

```html
# 资源包\Code\chapter3\3.8\4\flaskProject\templates\sell.html
<!DOCTYPE html>
<html lang = "en">
    <head>
        <meta charset = "UTF-8">
    </head>
    <body>
        销售订单页面<br>
        <p>销售订单页面的URL: {{ order_url }}</p>
    </body>
</html>
```

(8) 运行上述程序,打开浏览器并在网址栏输入 http://127.0.0.1:5000/,其显示内容如图 3-78 所示。

(9) 单击"个人设置页面"按钮,其显示的内容如图 3-79 所示。

图 3-78　浏览器中的显示内容 　　　　图 3-79　浏览器中的显示内容

(10) 单击"销售订单"按钮,其显示的内容如图 3-80 所示。

图 3-80　浏览器中的显示内容

3.8.5 在蓝图中实现子域名

在创建蓝图时,可以通过为 Blueprint 类添加 subdomain 属性实现子域名。

下面通过一个示例,演示一下如何在蓝图中实现子域名。

(1) 创建名为 flaskProject 的 Flask 项目。

(2) 打开保存在项目根目录下的文件 app.py,代码如下:

```
# 资源包\Code\chapter3\3.8\5\flaskProject\app.py
from flask import Flask, url_for, render_template
from blueprint.user import user
app = Flask(__name__)
# 用户模块
app.register_blueprint(blueprint = user)
@app.route('/')
def index():
    return render_template('index.html', user_url = url_for('user.settings'))
if __name__ == '__main__':
    app.run(debug = True)
```

(3) 在项目的根目录下的文件夹 templates 中创建模板文件 index.html,代码如下:

```
# 资源包\Code\chapter3\3.8\5\flaskProject\templates\index.html
<!DOCTYPE html>
<html lang = "en">
    <head>
        <meta charset = "UTF-8">
    </head>
    <body>
        <p>个人设置页面的 URL: {{ user_url }}</p>
    </body>
</html>
```

(4) 在项目的根目录下创建文件夹 blueprint,并在其中创建蓝图文件 user.py,代码如下:

```
# 资源包\Code\chapter3\3.8\5\flaskProject\blueprint\user.py
from flask import Blueprint
user = Blueprint('user', __name__, subdomain = 'user')
@user.route('/settings/')
def settings():
    return '个人设置页面'
```

(5) 运行上述程序,打开浏览器并在网址栏输入 http://user.127.0.0.1:5000/settings/,其结果是无法成功访问。

这是因为本地服务器 127.0.0.1 或 localhost 不支持子域名,并且由于其解析是通过本机的 C:\Windows\System32\drivers\etc\hosts 文件进行的,所以当在本地服务器测试子域名时需要修改 hosts 文件,并需要在应用程序 app.py 文件中配置子域名。

在文件 hosts 中添加以下域名与本机的映射。

```
127.0.0.1 oldxia86.com
127.0.0.1 user.oldxia86.com
```

修改保存在项目根目录下的文件 app.py，代码如下：

```python
from flask import Flask, url_for, render_template
from blueprint.user import user
app = Flask(__name__)
#配置子域名
app.config['SERVER_NAME'] = 'oldxia86.com:5000'
#用户模块
app.register_blueprint(blueprint = user)
@app.route('/')
def index():
    return render_template('index.html', user_url = url_for('user.settings'))
if __name__ == '__main__':
    app.run(debug = True)
```

（6）再次运行上述程序，打开浏览器并在网址栏输入 http://oldxia86.com:5000/，其显示内容如图 3-81 所示。

图 3-81　浏览器中的显示内容

（7）在浏览器中的网址栏输入 http://user.oldxia86.com:5000/settings/，其显示内容如图 3-82 所示。

图 3-82　浏览器中的显示内容

3.9　SQLAlchemy

SQLAlchemy 是 Python 中著名的 ORM 框架，其提供了 SQL 工具包及对象关系映射工具，可以简化应用程序开发人员在原生 SQL 上的操作，使开发人员可以将主要精力放在程序逻辑上，进而提高开发效率。此外，SQLAlchemy 还提供了一整套企业级持久性模式，用于高效和高性能进行数据库访问。

3.9.1　安装 SQLAlchemy

由于 SQLAlchemy 属于 Python 的第三方库，所以需要安装，只需在命令提示符中输入命令 pip install SQLAlchemy。

3.9.2　创建数据库引擎

可以通过 sqlalchemy 模块中的 create_engine()函数创建数据库引擎，并获得数据库引

擎对象,其语法格式如下:

```
create_engine(url)
```

其中,参数 url 表示数据库的 URL,其语法规则为 dialect[+driver]://username:password@host[:port]/databasename[?key=value..],dialect 表示数据库的类型,driver 表示数据库对应的驱动,username 表示连接数据库的用户名,password 表示连接数据库的密码,host 表示服务器地址,post 表示端口号,databasename 表示数据库的名称,key 和 value 表示传递数据库的参数,代码如下:

```python
#资源包\Code\chapter3\3.9\1\index.py
from sqlalchemy import create_engine
HOST = "127.0.0.1"
POST = "3306"
DATABASENAME = "flask_db"
USERNAME = "root"
PASSWORD = "12345678"
DB_URL = f"mysql+pymysql://{USERNAME}:{PASSWORD}@{HOST}:{POST}/{DATABASENAME}"
#创建数据库引擎
engine = create_engine(url=DB_URL)
```

3.9.3 创建数据库

创建数据库需要使用 sqlalchemy_utils 模块中的 create_database() 函数。由于该模块属于 Python 的第三方库,所以需要安装,在命令提示符中输入命令 pip install SQLAlchemy-Utils 即可。

需要注意的是,在创建数据库之前,需要使用 sqlalchemy_utils 模块中的 database_exists() 函数来判断待创建的数据库是否存在,代码如下:

```python
#资源包\Code\chapter3\3.9\2\index.py
from sqlalchemy import create_engine
from sqlalchemy_utils import create_database, database_exists
HOST = "127.0.0.1"
POST = "3306"
DATABASENAME = "flask_db"
USERNAME = "root"
PASSWORD = "12345678"
DB_URL = f"mysql+pymysql://{USERNAME}:{PASSWORD}@{HOST}:{POST}/{DATABASENAME}"
#创建数据库引擎
engine = create_engine(url=DB_URL)
#判断数据库是否存在
if not database_exists(DB_URL):
    create_database(DB_URL)
```

3.9.4 创建数据表

使用 SQLAlchemy 创建数据表依赖三大要素,即数据库引擎、基类和元素。

1. 数据库引擎

数据库引擎在之前的章节中已经详细讲解过,这里就不再赘述了。

2. 基类

在 ORM 中,数据模型通常为一个 Python 类,用于表示数据表,而该类的实例对象则表示数据表中的一条数据。

需要注意的是,该 Python 类必须继承自 sqlalchemy.orm 模块中的 declarative_base() 函数所创建的基类。

3. 元素

在 ORM 中,Python 类中的属性对应为数据表中的字段,而字段中的属性则可以通过 sqlalchemy 模块中的 Column 类创建,其语法格式如下:

```
Column(name, type_, primary_key, default, nullable, unique, autoincrement, onupdate)
```

其中:参数 name 用于指定属性映射到数据表中的字段名,如果不指定,则会使用这个属性的名字来作为字段名;参数 type_ 表示数据类型,而在 SQLAlchemy 中,需要通过 sqlalchemy 模块中的类来表示数据类型,其常用的数据类型如表 3-7 所示;参数 primary_key 用于设置字段是否为主键;参数 default 用于设置字段的默认值;参数 nullable 用于设置字段是否为空,其默认值为 True;参数 unique 用于设置字段的值是否唯一,即不允许有重复值,其默认值为 False;参数 autoincrement 用于设置当前字段是否为自增长;参数 onupdate 用于在数据更新时调用这个参数所指定的值或者函数。

表 3-7 常用的数据类型

数 据 类 型	描 述	数 据 类 型	描 述
Integer	整数	Time	时间
Float	浮点数	DateTime	日期和时间
Boolean	布尔值	String	字符类型
DECIMAL	定点类型	Text	文本类型
enum	枚举类型	LONGTEXT	长文本类型
Date	日期		

此外,在该类中还有一个特殊的属性__tablename__,用于表示数据表的名称。

最后,在创建完三大要素之后,可以通过基类的属性 metadata 创建 MetaData 对象,并使用 MetaData 对象的 create_all() 方法将数据模型映射到数据库中,以完成数据表的创建,其语法格式如下:

```
create_all(bind)
```

其中,参数 bind 表示数据库引擎。

这里需要注意的是,使用 create_all() 方法将数据模型映射到数据库中后,即使改变了数据模型的字段,也不会重新映射,所以,如果需要修改数据表的结构,则需要使用 MetaData 对象的 drop_all() 方法先删除旧的数据表,然后创建新的数据表。

下面通过一个示例,演示一下如何创建数据表。

(1) 创建主程序 index.py,代码如下:

```
#资源包\Code\chapter3\3.9\3\index.py
from sqlalchemy import create_engine, Column, Integer, String
```

```python
from sqlalchemy_utils import create_database, database_exists
from sqlalchemy.orm import declarative_base
HOST = "127.0.0.1"
POST = "3306"
DATABASENAME = "flask_db"
USERNAME = "root"
PASSWORD = "12345678"
DB_URL = f"mysql + pymysql://{USERNAME}:{PASSWORD}@{HOST}:{POST}/{DATABASENAME}"
#创建数据库引擎
engine = create_engine(url = DB_URL)
#判断数据库是否存在
if not database_exists(DB_URL):
    create_database(DB_URL)
#创建基类
Base = declarative_base()
class Person(Base):
    __tablename__ = 'person'
    id = Column(Integer, primary_key = True, autoincrement = True)
    name = Column(String(50))
    age = Column(Integer)
    country = Column(String(50))
MetaData = Base.metadata
MetaData.create_all(bind = engine)
```

(2) 运行上述程序。

(3) 打开命令提示符窗口,登入 MySQL。

(4) 在 MySQL 命令行窗口中输入 SQL 语句"use flask_db;",选择新创建的数据库 flask_db,其结果如图 3-83 所示。

(5) 输入 SQL 语句"show tables;",查询当前数据库中的数据表,其结果如图 3-84 所示。

(6) 输入 SQL 语句"desc person;",查询该数据表的表结构,其结果如图 3-85 所示。

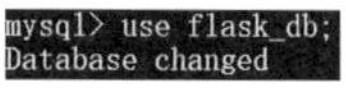

图 3-83 选择数据库    图 3-84 查询数据表    图 3-85 查看表结构

3.9.5 CRUD 操作

在 SQLAlchemy 中,首先需要通过 sqlalchemy.orm 模块中的 sessionmaker() 类创建会话类,其语法格式如下:

```
sessionmaker(bind)
```

其中,参数 bind 表示数据库引擎。

然后通过该类的相关方法,即可完成数据的插入、查询、更新和删除操作。

最后,通过会话类的 commit() 方法将会话中的所有操作作为一个事务提交至数据库,并使用 close()方法关闭会话。

1. 插入数据

通过会话类的 add()方法即可完成单条数据的添加操作,其语法格式如下:

```
add(instance)
```

其中,参数 instance 表示数据模型对象。

通过会话类的 add_all()方法即可完成多条数据的添加操作,其语法格式如下:

```
add_all(instances)
```

其中,参数 instances 表示数据模型对象组成的列表。

下面通过一个示例,演示一下如何插入数据。

(1) 创建主程序 index.py,代码如下:

```python
#资源包\Code\chapter3\3.9\4\index.py
from sqlalchemy import create_engine, Column, Integer, String
from sqlalchemy_utils import create_database, database_exists
from sqlalchemy.orm import declarative_base, sessionmaker
HOST = "127.0.0.1"
POST = "3306"
DATABASENAME = "flask_db"
USERNAME = "root"
PASSWORD = "12345678"
DB_URL = f"mysql+pymysql://{USERNAME}:{PASSWORD}@{HOST}:{POST}/{DATABASENAME}"
engine = create_engine(url=DB_URL)
if not database_exists(DB_URL):
    create_database(DB_URL)
Base = declarative_base()
class Person(Base):
    __tablename__ = 'person'
    id = Column(Integer, primary_key=True, autoincrement=True)
    name = Column(String(50))
    age = Column(Integer)
    country = Column(String(50))
MetaData = Base.metadata
MetaData.create_all(bind=engine)
Session = sessionmaker(bind=engine)
db_session = Session()
son = Person(name="xzd", age=35, country="China")
mother = Person(name="yp", age=68, country="China")
father = Person(name="xxg", age=68, country="China")
wife = Person(name="hxr", age=32, country=None)
db_session.add(son)
db_session.add_all([mother, father, wife])
db_session.commit()
db_session.close()
```

(2) 运行上述程序。

(3) 打开命令提示符窗口,登入 MySQL。

(4) 在 MySQL 命令行窗口中输入 SQL 语句"use flask_db;",选择新创建的数据库 flask_db,其结果如图 3-86 所示。

(5) 输入 SQL 语句 "select * from person;",查询数据表中的数据,其结果如图 3-87 所示。

图 3-86　选择数据库　　　　图 3-87　查询数据表中的数据

2. 查询数据

通过会话类的 query() 方法来获取 Query 查询对象,该对象的本质是将数据模型类映射为一条 SQL 语句,其语法格式如下:

```
query(entities)
```

其中,参数 entities 表示数据模型对象、数据模型对象的属性或聚合函数(该函数需要通过 sqlalchemy 模块中的 func 类的相关方法来实现,其常用的聚合函数如表 3-8 所示)。

表 3-8　常用的聚合函数

聚 合 函 数	描　　述
count()	统计行的数量
sum()	求和
max()	求最大值
min()	求最小值
avg()	求平均值

而在获取 Query 查询对象之后,就可以通过其相关方法(如表 3-9 所示)来获取查询结果。

表 3-9　Query 查询对象的相关方法

方　　法	描　　述
all()	返回由查询的全部结果的数据模型对象所组成的列表
first()	返回查询的第 1 个结果的数据模型对象
get()	根据主键 ID 获取数据模型对象,若主键不存在,则返回 None
count()	返回查询结果的数量

下面通过一个示例,演示一下如何查询数据。
(1) 创建主程序 index.py,代码如下:

```
#资源包\Code\chapter3\3.9\5\index.py
from sqlalchemy import create_engine, Column, Integer, String, func
from sqlalchemy_utils import create_database, database_exists
from sqlalchemy.orm import declarative_base, sessionmaker
HOST = "127.0.0.1"
POST = "3306"
DATABASENAME = "flask_db"
```

```python
USERNAME = "root"
PASSWORD = "12345678"
DB_URL = f"mysql+pymysql://{USERNAME}:{PASSWORD}@{HOST}:{POST}/{DATABASENAME}"
engine = create_engine(url=DB_URL)
if not database_exists(DB_URL):
    create_database(DB_URL)
Base = declarative_base()
class Person(Base):
    __tablename__ = 'person'
    id = Column(Integer, primary_key=True, autoincrement=True)
    name = Column(String(50))
    age = Column(Integer)
    country = Column(String(50))
MetaData = Base.metadata
MetaData.create_all(bind=engine)
Session = sessionmaker(bind=engine)
db_session = Session()
son = Person(name="xzd", age=35, country="China")
mother = Person(name="yp", age=68, country="China")
father = Person(name="xxg", age=68, country="China")
wife = Person(name="hxr", age=32, country=None)
db_session.add(son)
db_session.add_all([mother, father, wife])
db_session.commit()
query_person = db_session.query(Person)
print(query_person)
# 数据模型对象
all_person = query_person.all()
for per in all_person:
    print(per.name)
print('============ 数据模型对象 ================ ')
first_person = query_person.first()
print(first_person.name)
print('============ 数据模型对象 ================ ')
get_person = query_person.get(2)
print(get_person.name)
print('============ 数据模型对象 ================ ')
count_person = query_person.count()
print(count_person)
# 数据模型对象的属性
print('============ 数据模型对象的属性 ================ ')
query_person = db_session.query(Person.age)
print(query_person)
all_person = query_person.all()
print(all_person)
# 聚合函数
print('============ 聚合函数 ================ ')
query_person = db_session.query(func.count(Person.id))
print(query_person)
all_person = query_person.first()
print(all_person)
print('============ 聚合函数 ================ ')
query_person = db_session.query(func.avg(Person.age))
print(query_person)
all_person = query_person.first()
print(all_person)
db_session.commit()
db_session.close()
```

(2) 运行上述程序，PyCharm 中的运行结果如图 3-88 所示。

```
SELECT person.id AS person_id, person.name AS person_name, person.age AS person_age, person.country AS person_country
FROM person
xzd
yp
xxg
hxr
============数据模型对象================
xzd
============数据模型对象================
yp
============数据模型对象================
4
============数据模型对象的属性================
SELECT person.age AS person_age
FROM person
[(35,), (68,), (68,), (32,)]
============聚合函数================
SELECT count(person.id) AS count_1
FROM person
(4,)
============聚合函数================
SELECT avg(person.age) AS avg_1
FROM person
(Decimal('50.7500'),)
```

图 3-88　PyCharm 中的运行结果

除此之外，还可以通过查询条件过滤器 filter() 更精确地获取查询结果，需要注意的是，查询条件过滤器需要在查询方法之前使用，其语法格式如下：

filter(criterion)

其中，参数 criterion 表示筛选条件，其常用筛选条件如表 3-10 所示。

表 3-10　常用的筛选条件

筛 选 条 件	操作符或方法	筛 选 条 件	操作符或方法
等值过滤器	==	空值查询	is_()、isnot()
不等值过滤器	!=、<、>、<=、>=	逻辑非	~
模糊查询	like()	逻辑与	and_()
包括过滤器	in_()	逻辑或	or_()

下面通过一个示例，演示一下如何精确地查询数据。

(1) 创建主程序 index.py，代码如下：

```
# 资源包\Code\chapter3\3.9\6\index.py
from sqlalchemy import create_engine, Column, Integer, String, and_, or_
from sqlalchemy_utils import create_database, database_exists
from sqlalchemy.orm import declarative_base, sessionmaker
HOST = "127.0.0.1"
POST = "3306"
DATABASENAME = "flask_db"
USERNAME = "root"
PASSWORD = "12345678"
DB_URL = f"mysql+pymysql://{USERNAME}:{PASSWORD}@{HOST}:{POST}/{DATABASENAME}"
engine = create_engine(url=DB_URL)
if not database_exists(DB_URL):
    create_database(DB_URL)
```

```python
Base = declarative_base()
class Person(Base):
    __tablename__ = 'person'
    id = Column(Integer, primary_key = True, autoincrement = True)
    name = Column(String(50))
    age = Column(Integer)
    country = Column(String(50))
MetaData = Base.metadata
MetaData.create_all(bind = engine)
Session = sessionmaker(bind = engine)
db_session = Session()
son = Person(name = "xzd", age = 35, country = "China")
mother = Person(name = "yp", age = 68, country = "China")
father = Person(name = "xxg", age = 68, country = "China")
wife = Person(name = "hxr", age = 32, country = None)
db_session.add(son)
db_session.add_all([mother, father, wife])
db_session.commit()
print('============== id 等于 1 ===================== ')
query_person = db_session.query(Person).filter(Person.id == 1)
person_name = query_person.first()
print(person_name.name)
print('============== id 不等于 3 ===================== ')
query_person = db_session.query(Person).filter(Person.id != 3)
person_name = query_person.first()
print(person_name.name)
print('============== id 大于 1 ===================== ')
query_person = db_session.query(Person).filter(Person.id > 1)
person_name = query_person.first()
print(person_name.name)
print('============== id 小于 3 ==================== ')
query_person = db_session.query(Person).filter(Person.id < 3)
person_name = query_person.first()
print(person_name.name)
print('============== id 大于或等于 1 ===================== ')
query_person = db_session.query(Person).filter(Person.id >= 1)
person_name = query_person.first()
print(person_name.name)
print('============== id 小于或等于 3 ===================== ')
query_person = db_session.query(Person).filter(Person.id <= 3)
person_name = query_person.first()
print(person_name.name)
print('============== name 中包含 y ==================== ')
query_person = db_session.query(Person).filter(Person.name.like('%y%'))
person_name = query_person.first()
print(person_name.name)
print('============== id 在 1 和 2 中 ==================== ')
query_person = db_session.query(Person).filter(Person.id.in_([1, 2]))
person_names = query_person.first()
print('============== country 等于 NULL ==================== ')
query_person = db_session.query(Person).filter(Person.country.is_(None))
person_name = query_person.first()
print(person_name.name)
print('============== id 不等于 NULL ==================== ')
query_person = db_session.query(Person).filter(Person.id.isnot(None))
person_name = query_person.first()
```

```python
print(person_name.name)
print('================ id 不在 1 和 2 中 =================== ')
query_person = db_session.query(Person).filter(~Person.id.in_([1, 2]))
person_name = query_person.first()
print(person_name.name)
print('================ name 等于 yp,并且 age 等于 68 ================== ')
query_person = db_session.query(Person).filter(Person.name == 'yp', Person.age == 68)
person_name = query_person.first()
print(person_name.name)
print('================ name 等于 yp,并且 age 等于 68 ================== ')
query_person = db_session.query(Person).filter(and_(Person.name == 'yp', Person.age == 68))
person_name = query_person.first()
print(person_name.name)
print('================ name 等于 yp,并且 age 等于 68 ================== ')
query_person = db_session.query(Person).filter(Person.name == 'yp' and Person.age == 68)
person_name = query_person.first()
print(person_name.name)
print('================ name 等于 yp,并且 age 等于 68 ================== ')
query_person = db_session.query(Person).filter(Person.name == 'yp').filter(Person.age == 68)
person_name = query_person.first()
print(person_name.name)
print('================ name 等于 yp,或者 age 等于 68 ================== ')
query_person = db_session.query(Person).filter(or_(Person.name == 'yp', Person.age == 68))
person_name = query_person.first()
print(person_name.name)
db_session.commit()
db_session.close()
```

（2）运行上述程序,PyCharm 中的运行结果如图 3-89 所示。

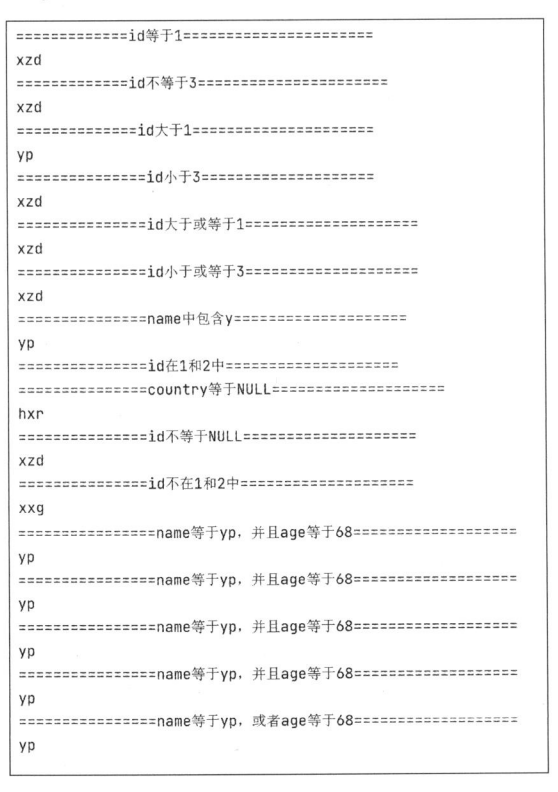

```
==============id等于1====================
xzd
==============id不等于3==================
xzd
==============id大于1====================
yp
==============id小于3====================
xzd
==============id大于或等于1===============
xzd
==============id小于或等于3===============
xzd
==============name中包含y================
yp
==============id在1和2中==================
==============country等于NULL=============
hxr
==============id不等于NULL================
xzd
==============id不在1和2中================
xxg
==============name等于yp,并且age等于68======
yp
==============name等于yp,并且age等于68======
yp
==============name等于yp,并且age等于68======
yp
==============name等于yp,并且age等于68======
yp
==============name等于yp,或者age等于68======
yp
```

图 3-89　PyCharm 中的运行结果

3. 删除数据

通过会话类的 delete()方法可以完成数据的删除操作，其语法格式如下：

```
delete(instance)
```

其中，参数 instance 表示待删除的数据模型对象。

下面通过一个示例，演示一下如何删除数据。

（1）创建主程序 index.py，代码如下：

```python
# 资源包\Code\chapter3\3.9\7\index.py
from sqlalchemy import create_engine, Column, Integer, String, and_, or_
from sqlalchemy_utils import create_database, database_exists
from sqlalchemy.orm import declarative_base, sessionmaker
HOST = "127.0.0.1"
POST = "3306"
DATABASENAME = "flask_db"
USERNAME = "root"
PASSWORD = "12345678"
DB_URL = f"mysql+pymysql://{USERNAME}:{PASSWORD}@{HOST}:{POST}/{DATABASENAME}"
engine = create_engine(url=DB_URL)
if not database_exists(DB_URL):
    create_database(DB_URL)
Base = declarative_base()
class Person(Base):
    __tablename__ = 'person'
    id = Column(Integer, primary_key=True, autoincrement=True)
    name = Column(String(50))
    age = Column(Integer)
    country = Column(String(50))
MetaData = Base.metadata
MetaData.create_all(bind=engine)
Session = sessionmaker(bind=engine)
db_session = Session()
son = Person(name="xzd", age=35, country="China")
mother = Person(name="yp", age=68, country="China")
father = Person(name="xxg", age=68, country="China")
wife = Person(name="hxr", age=32, country=None)
db_session.add(son)
db_session.add_all([mother, father, wife])
db_session.commit()
query_person = db_session.query(Person)
all_person = query_person.all()
print("删除前,数据库中的人名: ")
for person in all_person:
    print(person.name)
first_person = query_person.first()
db_session.delete(first_person)
print("删除后,数据库中的人名: ")
all_person = query_person.all()
for person in all_person:
    print(person.name)
db_session.commit()
db_session.close()
```

（2）运行上述程序，PyCharm 中的运行结果如图 3-90 所示。

```
删除前，数据库中的人名：
xzd
yp
xxg
hxr
删除后，数据库中的人名：
yp
xxg
hxr
```

图 3-90　PyCharm 中的运行结果

4. 修改数据

对数据库中的数据进行修改，只需修改查询出的数据模型对象的相关属性。

下面通过一个示例，演示一下如何修改数据。

（1）创建主程序 index.py，代码如下：

```python
#资源包\Code\chapter3\3.9\8\index.py
from sqlalchemy import create_engine, Column, Integer, String, and_, or_
from sqlalchemy_utils import create_database, database_exists
from sqlalchemy.orm import declarative_base, sessionmaker
HOST = "127.0.0.1"
POST = "3306"
DATABASENAME = "flask_db"
USERNAME = "root"
PASSWORD = "12345678"
DB_URL = f"mysql+pymysql://{USERNAME}:{PASSWORD}@{HOST}:{POST}/{DATABASENAME}"
engine = create_engine(url=DB_URL)
if not database_exists(DB_URL):
    create_database(DB_URL)
Base = declarative_base()
class Person(Base):
    __tablename__ = 'person'
    id = Column(Integer, primary_key=True, autoincrement=True)
    name = Column(String(50))
    age = Column(Integer)
    country = Column(String(50))
MetaData = Base.metadata
MetaData.create_all(bind=engine)
Session = sessionmaker(bind=engine)
db_session = Session()
son = Person(name="xzd", age=35, country="China")
mother = Person(name="yp", age=68, country="China")
father = Person(name="xxg", age=68, country="China")
wife = Person(name="hxr", age=32, country=None)
db_session.add(son)
db_session.add_all([mother, father, wife])
db_session.commit()
query_person = db_session.query(Person)
first_person = query_person.first()
print(f'修改前的姓名：{first_person.name}')
first_person.name = '于萍'
first_person = query_person.first()
print(f'修改后的姓名：{first_person.name}')
db_session.commit()
db_session.close()
```

（2）运行上述程序，PyCharm 中的运行结果如图 3-91 所示。

修改前的姓名：yp
修改后的姓名：于萍

图 3-91　PyCharm 中的运行结果

3.9.6　外键

在 MySQL 中，外键可以让多表之间的关系更加紧密，而 SQLAlchemy 同样支持外键，其通过 ForeignKey 类来实现，其语法格式如下：

```
ForeignKey(column, ondelete)
```

其中：参数 column 表示主表中的主键；参数 ondelete 用于设置外键约束，包括 RESTRICT（默认约束，表示当删除父表数据时，如果子表有数据在使用该字段中的数据，则会阻止删除）、NO ACTION（同 RESTRICT）、CASCADE（表示级联删除，即删除父表的某一条数据时，其子表中使用该外键的数据也会被删除）和 SET NULL（表示当删除父表中的某一条数据时，其子表中使用该外键的数据设置为 NULL）。

下面通过一个示例，演示一下如何使用外键。

（1）创建主程序 index.py，代码如下：

```python
# 资源包\Code\chapter3\3.9\9\index.py
from sqlalchemy import create_engine, Column, Integer, String, Text, ForeignKey
from sqlalchemy_utils import create_database, database_exists
from sqlalchemy.orm import declarative_base, sessionmaker
HOST = "127.0.0.1"
POST = "3306"
DATABASENAME = "flask_db"
USERNAME = "root"
PASSWORD = "12345678"
DB_URL = f"mysql+pymysql://{USERNAME}:{PASSWORD}@{HOST}:{POST}/{DATABASENAME}"
engine = create_engine(url=DB_URL)
if not database_exists(DB_URL):
    create_database(DB_URL)
Base = declarative_base()
class User(Base):
    __tablename__ = 'user'
    id = Column(Integer, primary_key=True, autoincrement=True)
    username = Column(String(50), nullable=False)
class Article(Base):
    __tablename__ = "article"
    id = Column(Integer, primary_key=True, autoincrement=True)
    title = Column(String(50), nullable=False)
    content = Column(Text, nullable=False)
    # 外键约束，默认为 RESTRICT
    uid = Column(Integer, ForeignKey("user.id"))
    # uid = Column(Integer, ForeignKey("user.id", ondelete="RESTRICT"))
    # uid = Column(Integer, ForeignKey("user.id", ondelete="NO ACTION"))
    # uid = Column(Integer, ForeignKey("user.id", ondelete="CASCADE"))
    # uid = Column(Integer, ForeignKey("user.id", ondelete="SET NULL"))
MetaData = Base.metadata
```

```
MetaData.drop_all(bind = engine)
MetaData.create_all(bind = engine)
Session = sessionmaker(bind = engine)
db_session = Session()
user = User(username = 'xzd')
db_session.add(user)
db_session.commit()
article = Article(title = '《Python 全栈开发——Web 编程》', content = '《Python 全栈开发——Web
编程》', uid = 1)
db_session.add(article)
db_session.commit()
db_session.close()
```

(2) 运行上述程序。

(3) 打开命令提示符窗口,登入 MySQL。

(4) 在 MySQL 命令行窗口中输入 SQL 语句“delete from user where id=1;”,此时,不同的外键约束会显示不同的结果:

一是当外键约束为 RESTRICT 时,其结果如图 3-92 所示。

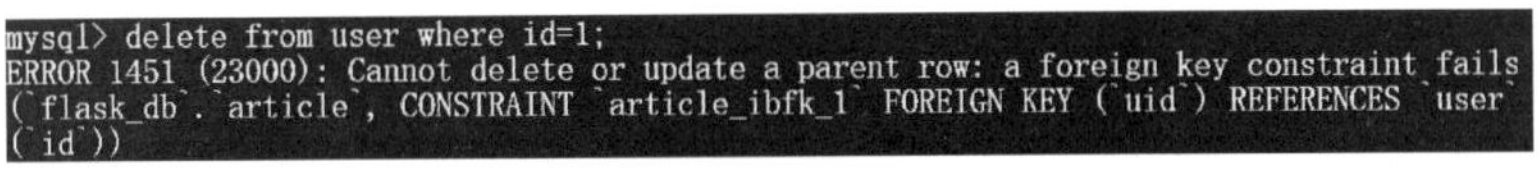

```
mysql> delete from user where id=1;
ERROR 1451 (23000): Cannot delete or update a parent row: a foreign key constraint fails
(`flask_db`.`article`, CONSTRAINT `article_ibfk_1` FOREIGN KEY (`uid`) REFERENCES `user`
(`id`))
```

图 3-92 外键约束为 RESTRICT

二是当外键约束为 NO ACTION 时,其结果如图 3-93 所示。

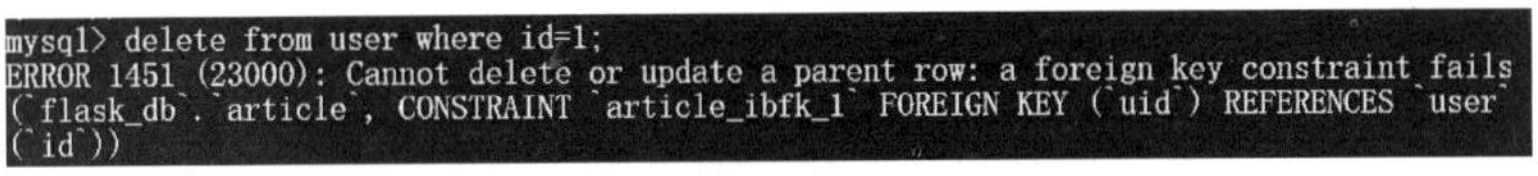

```
mysql> delete from user where id=1;
ERROR 1451 (23000): Cannot delete or update a parent row: a foreign key constraint fails
(`flask_db`.`article`, CONSTRAINT `article_ibfk_1` FOREIGN KEY (`uid`) REFERENCES `user`
(`id`))
```

图 3-93 外键约束为 NO ACTION

三是当外键约束为 CASCADE 时,其结果如图 3-94 所示。

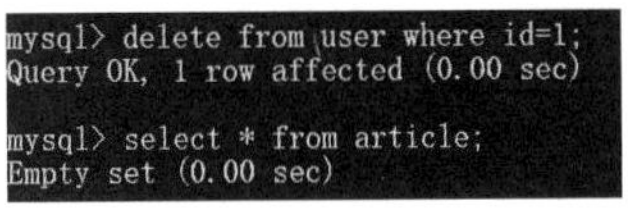

```
mysql> delete from user where id=1;
Query OK, 1 row affected (0.00 sec)

mysql> select * from article;
Empty set (0.00 sec)
```

图 3-94 外键约束为 CASCADE

四是当外键约束为 SET NULL 时,其结果如图 3-95 所示。

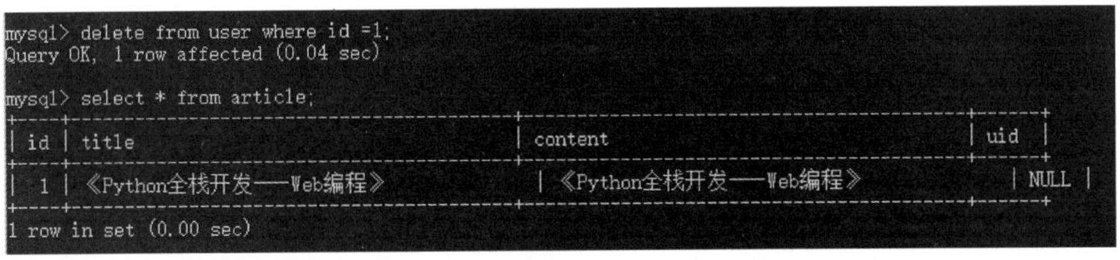

```
mysql> delete from user where id =1;
Query OK, 1 row affected (0.04 sec)

mysql> select * from article;
+----+------------------------------+------------------------------+------+
| id | title                        | content                      | uid  |
+----+------------------------------+------------------------------+------+
|  1 | 《Python全栈开发——Web编程》  | 《Python全栈开发——Web编程》  | NULL |
+----+------------------------------+------------------------------+------+
1 row in set (0.00 sec)
```

图 3-95 外键约束为 SET NULL

3.9.7 多表间关系

在数据结构上,外键对连表的查询并没有太多的帮助,但在 SQLAlchemy 中,外键对连表的查询进行了一定的优化,即可通过与sqlalchemy.orm模块中的relationship()函数进行搭配使用以处理一对多关系、一对一关系、多对多关系或多对一关系,其语法格式如下:

> relationship(argument, backref, uselist, secondary, lazy, cascade)

其中:参数argument表示关系另一侧的数据模型名称;参数backref表示对关系提供反向引用的声明;参数uselist用于控制其返回的是列表还是单个对象,当其值为False时,一对多关系将被转换为一对一关系;参数secondary用于在多对多关系中指定关联表;参数lazy表示为多表关联而定义的一系列加载方法,常用的方法包括select(默认方法,即一次性加载全部数据,并返回列表)和dynamic(不直接加载数据,而返回一个AppenderQuery对象,该对象与Query查询对象一样,可以对该对象再进行一层过滤和排序等操作);参数cascade用于设置级联操作。

1. 一对多关系

一对多关系表示一个数据模型可以对应多个其他数据模型,而其他数据模型只能对应一个数据模型,例如,有两个数据模型User和Article,即一个用户可以编写多篇文章,而一篇文章只能有一个用户。

下面通过一个示例,演示一下如何实现一对多关系。

(1) 创建主程序index.py,代码如下:

```python
# 资源包\Code\chapter3\3.9\10\index.py
from sqlalchemy import create_engine, Column, Integer, String, Text, ForeignKey
from sqlalchemy_utils import create_database, database_exists
from sqlalchemy.orm import declarative_base, sessionmaker, relationship
HOST = "127.0.0.1"
POST = "3306"
DATABASENAME = "flask_db"
USERNAME = "root"
PASSWORD = "12345678"
DB_URL = f"mysql+pymysql://{USERNAME}:{PASSWORD}@{HOST}:{POST}/{DATABASENAME}"
engine = create_engine(url=DB_URL)
if not database_exists(DB_URL):
    create_database(DB_URL)
Base = declarative_base()
class User(Base):
    __tablename__ = 'user'
    id = Column(Integer, primary_key=True, autoincrement=True)
    username = Column(String(50), nullable=False)
    articles = relationship(argument="Article")
class Article(Base):
    __tablename__ = "article"
    id = Column(Integer, primary_key=True, autoincrement=True)
    title = Column(String(50), nullable=False)
    content = Column(Text, nullable=False)
    uid = Column(Integer, ForeignKey("user.id"))
    author = relationship("User")
MetaData = Base.metadata
MetaData.drop_all(bind=engine)
MetaData.create_all(bind=engine)
Session = sessionmaker(bind=engine)
```

```
db_session = Session()
user = User(username = '夏正东')
db_session.add(user)
db_session.commit()
article1 = Article(title = '《Python全栈开发——Web编程》', content = '《Python全栈开发——Web编程》', uid = 1)
article2 = Article(title = '《Python全栈开发——数据分析》', content = '《Python全栈开发——数据分析》', uid = 1)
article3 = Article(title = '《Python全栈开发——高阶编程》', content = '《Python全栈开发——高阶编程》', uid = 1)
db_session.add_all([article1, article2, article3])
article = db_session.query(Article).first()
print(article.author.username)
user = db_session.query(User).first()
for article in user.articles:
    print(article.title)
db_session.commit()
db_session.close()
```

（2）运行上述程序，PyCharm中的运行结果如图3-96所示。

此外，除了可以在两个数据模型中分别使用relationship()函数进行关联，还可以通过relationship()函数中的参数backref进行反向关联。

```
夏正东
《Python全栈开发——Web编程》
《Python全栈开发——数据分析》
《Python全栈开发——高阶编程》
```

图3-96　PyCharm中的运行结果

下面通过一个示例，演示一下如何实现反向关联的一对多关系。

（1）创建主程序index.py，代码如下：

```
# 资源包\Code\chapter3\3.9\11\index.py
from sqlalchemy import create_engine, Column, Integer, String, Text, ForeignKey
from sqlalchemy_utils import create_database, database_exists
from sqlalchemy.orm import declarative_base, sessionmaker, relationship
HOST = "127.0.0.1"
POST = "3306"
DATABASENAME = "flask_db"
USERNAME = "root"
PASSWORD = "12345678"
DB_URL = f"mysql+pymysql://{USERNAME}:{PASSWORD}@{HOST}:{POST}/{DATABASENAME}"
engine = create_engine(url = DB_URL)
if not database_exists(DB_URL):
    create_database(DB_URL)
Base = declarative_base()
class User(Base):
    __tablename__ = 'user'
    id = Column(Integer, primary_key = True, autoincrement = True)
    username = Column(String(50), nullable = False)
class Article(Base):
    __tablename__ = "article"
    id = Column(Integer, primary_key = True, autoincrement = True)
    title = Column(String(50), nullable = False)
    content = Column(Text, nullable = False)
    uid = Column(Integer, ForeignKey("user.id"))
    author = relationship(argument = "User", backref = "articles")
```

```
MetaData = Base.metadata
MetaData.drop_all(bind = engine)
MetaData.create_all(bind = engine)
Session = sessionmaker(bind = engine)
db_session = Session()
user = User(username = '夏正东')
db_session.add(user)
db_session.commit()
article1 = Article(title = '《Python全栈开发——Web编程》', content = '《Python全栈开发——Web编程》', uid = 1)
article2 = Article(title = '《Python全栈开发——数据分析》', content = '《Python全栈开发——数据分析》', uid = 1)
article3 = Article(title = '《Python全栈开发——高阶编程》', content = '《Python全栈开发——高阶编程》', uid = 1)
db_session.add_all([article1, article2, article3])
article = db_session.query(Article).first()
print(article.author.username)
user = db_session.query(User).first()
for article in user.articles:
    print(article.title)
db_session.commit()
db_session.close()
```

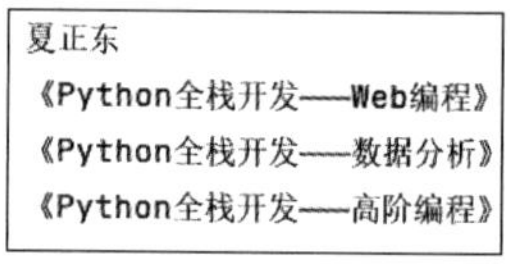

夏正东
《Python全栈开发——Web编程》
《Python全栈开发——数据分析》
《Python全栈开发——高阶编程》

图 3-97 PyCharm 中的运行结果

(2) 运行上述程序,PyCharm 中的运行结果如图 3-97 所示。

2. 一对一关系

一对一关系表示两个数据模型之间存在一个唯一的对应关系,例如,有两个数据模型 User 和 UserExtend,即一个用户只能有一个该用户的用户扩展,而用户扩展也只能对应一个用户。

下面通过一个示例,演示一下如何实现一对一关系。

(1) 创建主程序 index.py,代码如下:

```
# 资源包\Code\chapter3\3.9\12\index.py
from sqlalchemy import create_engine, Column, Integer, String, Text, ForeignKey
from sqlalchemy_utils import create_database, database_exists
from sqlalchemy.orm import declarative_base, sessionmaker, relationship
HOST = "127.0.0.1"
POST = "3306"
DATABASENAME = "flask_db"
USERNAME = "root"
PASSWORD = "12345678"
DB_URL = f"mysql+pymysql://{USERNAME}:{PASSWORD}@{HOST}:{POST}/{DATABASENAME}"
engine = create_engine(url = DB_URL)
if not database_exists(DB_URL):
    create_database(DB_URL)
Base = declarative_base()
class User(Base):
    __tablename__ = 'user'
    id = Column(Integer, primary_key = True, autoincrement = True)
    username = Column(String(50), nullable = False)
    extend = relationship("UserExtend", uselist = False)
class UserExtend(Base):
    __tablename__ = 'user_extend'
```

```
    id = Column(Integer, primary_key = True, autoincrement = True)
    job = Column(String(50))
    uid = Column(Integer, ForeignKey("user.id"))
    user = relationship("User")
    #user = relationship("User",backref = backref("extend",uselist = False))
class Article(Base):
    __tablename__ = "article"
    id = Column(Integer, primary_key = True, autoincrement = True)
    title = Column(String(50), nullable = False)
    content = Column(Text, nullable = False)
    uid = Column(Integer, ForeignKey("user.id"))
    author = relationship(argument = "User", backref = "articles")
MetaData = Base.metadata
MetaData.drop_all(bind = engine)
MetaData.create_all(bind = engine)
Session = sessionmaker(bind = engine)
db_session = Session()
user = User(username = '夏正东')
use_extend = UserExtend(job = 'Teacher')
user.extend = use_extend
db_session.add(user)
db_session.commit()
db_session.close()
```

(2) 运行上述程序。

(3) 打开命令提示符窗口,登入 MySQL。

(4) 在 MySQL 命令行窗口中输入 SQL 语句"select * from user;",其结果如图 3-98 所示。

(5) 输入 SQL 语句"select * from user_extend;",其结果如图 3-99 所示。

图 3-98　查询数据表中的数据　　　图 3-99　查询数据表中的数据

3. 多对多关系

多对多关系表示两个数据模型之间存在多个对应关系,例如,有两个数据模型 Article 和 Tag,即一篇文章可以对应多个标签,并且一个标签也可以对应多篇文章。

下面通过一个示例,演示一下如何实现多对多关系。

(1) 创建主程序 index.py,代码如下:

```
#资源包\Code\chapter3\3.9\13\index.py
from sqlalchemy import create_engine, Column, Integer, String, ForeignKey, Table
from sqlalchemy_utils import create_database, database_exists
from sqlalchemy.orm import declarative_base, sessionmaker, relationship
HOST = "127.0.0.1"
POST = "3306"
DATABASENAME = "flask_db"
USERNAME = "root"
```

```python
    PASSWORD = "12345678"
    DB_URL = f"mysql+pymysql://{USERNAME}:{PASSWORD}@{HOST}:{POST}/{DATABASENAME}"
engine = create_engine(url=DB_URL)
if not database_exists(DB_URL):
    create_database(DB_URL)
Base = declarative_base()
# 关联表
article_tag = Table(
    "article_tag",
    Base.metadata,
    Column("article_id", Integer, ForeignKey("article.id"), primary_key=True),
    Column("tag_id", Integer, ForeignKey("tag.id"), primary_key=True)
)
class Article(Base):
    __tablename__ = "article"
    id = Column(Integer, primary_key=True, autoincrement=True)
    title = Column(String(50), nullable=False)
    tags = relationship('Tag', backref='articles', secondary=article_tag)
class Tag(Base):
    __tablename__ = "tag"
    id = Column(Integer, primary_key=True, autoincrement=True)
    name = Column(String(50), nullable=False)
    # articles = relationship("Article", backref="tags", secondary=article_tag)
MetaData = Base.metadata
MetaData.drop_all(bind=engine)
MetaData.create_all(bind=engine)
Session = sessionmaker(bind=engine)
db_session = Session()
article1 = Article(title="Python全栈开发——数据分析")
article2 = Article(title="Python全栈开发——Web编程")
tag1 = Tag(name="Python")
tag2 = Tag(name="夏正东")
article1.tags.append(tag1)
article1.tags.append(tag2)
article2.tags.append(tag1)
article2.tags.append(tag2)
db_session.add(article1)
db_session.add(article2)
db_session.commit()
db_session.close()
article = db_session.query(Article).first()
for tag in article.tags:
    print(tag.name)
print('==============')
tag = db_session.query(Tag).first()
for article in tag.articles:
    print(article.title)
```

(2) 运行上述程序，PyCharm 中的运行结果如图 3-100 所示。

在上述代码中，由于关联表 article_tag 并不是数据模型，而只是一个简单的数据表，所以只需利用 Table 类进行创建，其语法格式如下：

```
Python
夏正东
==============
Python全栈开发——数据分析
Python全栈开发——Web编程
```

图 3-100　PyCharm 中的运行结果

```
Table(name, metadata, column_list)
```

其中,参数 name 表示数据表名,参数 metadata 表示 MetaData 对象,参数 column_list 表示 Column 类。

3.9.8 高级查询

1. 排序查询

SQLAlchemy 实现排序的方式包括两种。

1) 使用 order_by() 方法

该方法可以根据数据模型中的指定属性实现正序排序,而通过"数据模型名称.属性名.desc()"或在数据模型中的指定属性前添加"-"可以实现降序排序。

下面通过一个示例,演示一下如何进行排序查询。

(1) 创建主程序 index.py,代码如下:

```python
# 资源包\Code\chapter3\3.9\14\index.py
from sqlalchemy import create_engine, Column, Integer, String, DateTime
from sqlalchemy_utils import create_database, database_exists
from sqlalchemy.orm import declarative_base, sessionmaker
from datetime import datetime
import time
HOST = "127.0.0.1"
POST = "3306"
DATABASENAME = "flask_db"
USERNAME = "root"
PASSWORD = "12345678"
DB_URL = f"mysql+pymysql://{USERNAME}:{PASSWORD}@{HOST}:{POST}/{DATABASENAME}"
engine = create_engine(url=DB_URL)
if not database_exists(DB_URL):
    create_database(DB_URL)
Base = declarative_base()
class Article(Base):
    __tablename__ = "article"
    id = Column(Integer, primary_key=True, autoincrement=True)
    title = Column(String(50), nullable=False)
    create_time = Column(DateTime, nullable=False, default=datetime.now)
MetaData = Base.metadata
MetaData.drop_all(bind=engine)
MetaData.create_all(bind=engine)
Session = sessionmaker(bind=engine)
db_session = Session()
article1 = Article(title='《Python 全栈开发——Web 编程》')
article2 = Article(title='《Python 全栈开发——数据分析》')
article3 = Article(title='《Python 全栈开发——高阶编程》')
db_session.add(article1)
db_session.commit()
time.sleep(1)
db_session.add(article2)
db_session.commit()
time.sleep(1)
db_session.add(article3)
db_session.commit()
articles = db_session.query(Article).order_by(Article.create_time).all()
# 降序排序
# articles = db_session.query(Article).order_by(Article.create_time.desc()).all()
```

```
# 降序排序
# articles = db_session.query(Article).order_by(-Article.create_time).all()
for article in articles:
    print(f'{article.title} - {article.create_time}')
db_session.commit()
db_session.close()
```

(2) 运行上述程序，PyCharm 中的运行结果如图 3-101 所示。

```
《Python全栈开发——Web编程》-2024-10-23 15:45:04
《Python全栈开发——数据分析》-2024-10-23 15:45:05
《Python全栈开发——高阶编程》-2024-10-23 15:45:06
```

图 3-101　PyCharm 中的运行结果

2) 使用 backref() 函数中的参数 order_by

在一对多关系中，可以通过 backref() 函数中的参数 order_by 进行排序。

下面通过一个示例，演示一下如何进行排序查询。

(1) 创建主程序 index.py，代码如下：

```
# 资源包\Code\chapter3\3.9\15\index.py
from sqlalchemy import create_engine, Column, Integer, String, ForeignKey, DateTime
from sqlalchemy_utils import create_database, database_exists
from sqlalchemy.orm import declarative_base, sessionmaker, relationship, backref
from datetime import datetime
import time
HOST = "127.0.0.1"
POST = "3306"
DATABASENAME = "flask_db"
USERNAME = "root"
PASSWORD = "12345678"
DB_URL = f"mysql+pymysql://{USERNAME}:{PASSWORD}@{HOST}:{POST}/{DATABASENAME}"
engine = create_engine(url=DB_URL)
if not database_exists(DB_URL):
    create_database(DB_URL)
Base = declarative_base()
class User(Base):
    __tablename__ = "user"
    id = Column(Integer, primary_key=True, autoincrement=True)
    username = Column(String(50), nullable=False)
class Article(Base):
    __tablename__ = "article"
    id = Column(Integer, primary_key=True, autoincrement=True)
    title = Column(String(50), nullable=False)
    create_time = Column(DateTime, nullable=False, default=datetime.now)
    uid = Column(Integer, ForeignKey('user.id'))
    author = relationship("User", backref=backref("articles", order_by=create_time))
MetaData = Base.metadata
MetaData.drop_all(bind=engine)
MetaData.create_all(bind=engine)
Session = sessionmaker(bind=engine)
db_session = Session()
article1 = Article(title='《Python全栈开发——Web编程》')
article2 = Article(title='《Python全栈开发——数据分析》')
```

```python
user = User(username = '夏正东')
user.articles = [article1, article2]
db_session.add(user)
db_session.commit()
time.sleep(2)
article3 = Article(title = '《Python全栈开发——高阶编程》')
user.articles.append(article3)
db_session.commit()
user = db_session.query(User).first()
for article in user.articles:
    print(f'{article.title} - {article.create_time}')
db_session.commit()
db_session.close()
```

（2）运行上述程序，PyCharm中的运行结果如图3-102所示。

```
《Python全栈开发——Web编程》-2024-10-23 15:46:18
《Python全栈开发——数据分析》-2024-10-23 15:46:18
《Python全栈开发——高阶编程》-2024-10-23 15:46:20
```

图 3-102　PyCharm 中的运行结果

2. 限量查询

通过limit()方法可以限制查询结果返回的数量。

下面通过示例演示一下如何进行限量查询。

（1）创建主程序index.py，代码如下：

```python
#资源包\Code\chapter3\3.9\16\index.py
from sqlalchemy import create_engine, Column, Integer, String
from sqlalchemy_utils import create_database, database_exists
from sqlalchemy.orm import declarative_base, sessionmaker
HOST = "127.0.0.1"
POST = "3306"
DATABASENAME = "flask_db"
USERNAME = "root"
PASSWORD = "12345678"
DB_URL = f"mysql + pymysql://{USERNAME}:{PASSWORD}@{HOST}:{POST}/{DATABASENAME}"
engine = create_engine(url = DB_URL)
if not database_exists(DB_URL):
    create_database(DB_URL)
Base = declarative_base()
class Article(Base):
    __tablename__ = "article"
    id = Column(Integer, primary_key = True, autoincrement = True)
    title = Column(String(50), nullable = False)
MetaData = Base.metadata
MetaData.drop_all(bind = engine)
MetaData.create_all(bind = engine)
Session = sessionmaker(bind = engine)
db_session = Session()
for i in range(50):
    title = f"title - {i}"
    article = Article(title = title)
    db_session.add(article)
```

```
    db_session.commit()
articles = db_session.query(Article).limit(10).all()
for article in articles:
    print(article.title)
```

（2）运行上述程序，PyCharm 中的运行结果如图 3-103 所示。

```
title-0
title-1
title-2
title-3
title-4
title-5
title-6
title-7
title-8
title-9
```

图 3-103　PyCharm 中的运行结果

3. 偏移查询

通过 offset()方法可以控制查询结果的偏移量。

下面通过示例演示一下如何进行偏移查询。

（1）创建主程序 index.py，代码如下：

```
# 资源包\Code\chapter3\3.9\17\index.py
from sqlalchemy import create_engine, Column, Integer, String
from sqlalchemy_utils import create_database, database_exists
from sqlalchemy.orm import declarative_base, sessionmaker
HOST = "127.0.0.1"
POST = "3306"
DATABASENAME = "flask_db"
USERNAME = "root"
PASSWORD = "12345678"
DB_URL = f"mysql+pymysql://{USERNAME}:{PASSWORD}@{HOST}:{POST}/{DATABASENAME}"
engine = create_engine(url=DB_URL)
if not database_exists(DB_URL):
    create_database(DB_URL)
Base = declarative_base()
class Article(Base):
    __tablename__ = "article"
    id = Column(Integer, primary_key=True, autoincrement=True)
    title = Column(String(50), nullable=False)
MetaData = Base.metadata
MetaData.drop_all(bind=engine)
MetaData.create_all(bind=engine)
Session = sessionmaker(bind=engine)
db_session = Session()
for i in range(50):
    title = f"title-{i}"
    article = Article(title=title)
    db_session.add(article)
    db_session.commit()
articles = db_session.query(Article).offset(5).limit(10).all()
for article in articles:
    print(article.title)
```

（2）运行上述程序，PyCharm 中的运行结果如图 3-104 所示。

```
title-5
title-6
title-7
title-8
title-9
title-10
title-11
title-12
title-13
title-14
```

图 3-104　PyCharm 中的运行结果

4．切片查询

通过 slice()方法可以对 Query 查询对象进行切片操作，以获取指定的数据。此外，也可以直接使用"[start:stop]"的形式进行切片操作。

下面通过一个示例演示一下如何进行切片查询。

（1）创建主程序 index.py，代码如下：

```python
# 资源包\Code\chapter3\3.9\18\index.py
from sqlalchemy import create_engine, Column, Integer, String
from sqlalchemy_utils import create_database, database_exists
from sqlalchemy.orm import declarative_base, sessionmaker
HOST = "127.0.0.1"
POST = "3306"
DATABASENAME = "flask_db"
USERNAME = "root"
PASSWORD = "12345678"
DB_URL = f"mysql+pymysql://{USERNAME}:{PASSWORD}@{HOST}:{POST}/{DATABASENAME}"
engine = create_engine(url=DB_URL)
if not database_exists(DB_URL):
    create_database(DB_URL)
Base = declarative_base()
class Article(Base):
    __tablename__ = "article"
    id = Column(Integer, primary_key=True, autoincrement=True)
    title = Column(String(50), nullable=False)
MetaData = Base.metadata
MetaData.drop_all(bind=engine)
MetaData.create_all(bind=engine)
Session = sessionmaker(bind=engine)
db_session = Session()
for i in range(50):
    title = f"title-{i}"
    article = Article(title=title)
    db_session.add(article)
    db_session.commit()
articles = db_session.query(Article).slice(5,15).all()
# articles = db_session.query(Article)[6:16]
for article in articles:
    print(article.title)
```

（2）运行上述程序，PyCharm 中的运行结果如图 3-105 所示。

```
title-5
title-6
title-7
title-8
title-9
title-10
title-11
title-12
title-13
title-14
```

图 3-105　PyCharm 中的运行结果

5．懒加载查询

懒加载查询指的是在一对多或者多对多关系的情况下，只获取一部分数据，而不是获取全部数据时所使用的查询技术。

可以通过给 relationship()方法添加属性 lazy 实现懒加载查询。

下面通过一个示例演示一下如何进行懒加载查询。

（1）创建主程序 index.py，代码如下：

```python
# 资源包\Code\chapter3\3.9\19\index.py
from sqlalchemy import create_engine, Column, Integer, String, ForeignKey
from sqlalchemy_utils import create_database, database_exists
from sqlalchemy.orm import declarative_base, sessionmaker, relationship, backref
HOST = "127.0.0.1"
POST = "3306"
DATABASENAME = "flask_db"
USERNAME = "root"
PASSWORD = "12345678"
DB_URL = f"mysql+pymysql://{USERNAME}:{PASSWORD}@{HOST}:{POST}/{DATABASENAME}"
engine = create_engine(url=DB_URL)
if not database_exists(DB_URL):
    create_database(DB_URL)
Base = declarative_base()
class User(Base):
    __tablename__ = "user"
    id = Column(Integer, primary_key=True, autoincrement=True)
    username = Column(String(50), nullable=False)
class Article(Base):
    __tablename__ = "article"
    id = Column(Integer, primary_key=True, autoincrement=True)
    title = Column(String(50), nullable=False)
    uid = Column(Integer, ForeignKey('user.id'))
    author = relationship("User", backref=backref("articles", lazy='dynamic'))
MetaData = Base.metadata
MetaData.drop_all(bind=engine)
MetaData.create_all(bind=engine)
Session = sessionmaker(bind=engine)
db_session = Session()
user = User(username='夏正东')
for i in range(50):
    title = f"title-{i}"
    article = Article(title=title)
    article.author = user
    db_session.add(article)
```

```
    db_session.commit()
user = db_session.query(User).first()
print(type(user.articles))
print('====================== ')
print(user.articles.all())
print('====================== ')
print(user.articles.filter(Article.id > 20).all())
print('====================== ')
article = Article(title = "title - 100")
user.articles.append(article)
db_session.commit()
```

(2) 运行上述程序，PyCharm 中的运行结果如图 3-106 所示。

```
<class 'sqlalchemy.orm.dynamic.AppenderQuery'>
======================
[<__main__.Article object at 0x046D7D90>, <__main__.Article object at 0x046D7D70>, <__main__.Article object at 0x046D7D50>, <__main__
.Article object at 0x046D7D30>, <__main__.Article object at 0x046D7D00>, <__main__.Article object at 0x046D7E10>, <__main__.Article
object at 0x046D7E30>, <__main__.Article object at 0x046D7E50>, <__main__.Article object at 0x046D7E70>, <__main__.Article object at
0x046D7DF0>, <__main__.Article object at 0x046D7E90>, <__main__.Article object at 0x046D7EB0>, <__main__.Article object at
0x046D7ED0>, <__main__.Article object at 0x046D7EF0>, <__main__.Article object at 0x046D7F10>, <__main__.Article object at
0x046D7F30>, <__main__.Article object at 0x046D7F50>, <__main__.Article object at 0x046D7F70>, <__main__.Article object at
0x046D7F90>, <__main__.Article object at 0x046D7FB0>, <__main__.Article object at 0x046D7FD0>, <__main__.Article object at
0x046D7FF0>, <__main__.Article object at 0x046E6030>, <__main__.Article object at 0x046E6050>, <__main__.Article object at
0x046E6070>, <__main__.Article object at 0x046E6090>, <__main__.Article object at 0x046E60B0>, <__main__.Article object at
0x046E60D0>, <__main__.Article object at 0x046E60F0>, <__main__.Article object at 0x046E6110>, <__main__.Article object at
0x046E6130>, <__main__.Article object at 0x046E6150>, <__main__.Article object at 0x046E6170>, <__main__.Article object at
0x046E6190>, <__main__.Article object at 0x046E61B0>, <__main__.Article object at 0x046E61D0>, <__main__.Article object at
0x046E61F0>, <__main__.Article object at 0x046E6210>, <__main__.Article object at 0x046E6230>, <__main__.Article object at
0x046E6250>, <__main__.Article object at 0x046E6270>, <__main__.Article object at 0x046E6290>, <__main__.Article object at
0x046E62B0>, <__main__.Article object at 0x046E62D0>, <__main__.Article object at 0x046E62F0>, <__main__.Article object at
0x046E6310>, <__main__.Article object at 0x046E6330>, <__main__.Article object at 0x046E6350>, <__main__.Article object at
0x046E6370>, <__main__.Article object at 0x046A5610>]
======================
[<__main__.Article object at 0x046E60F0>, <__main__.Article object at 0x046E60D0>, <__main__.Article object at 0x046E60B0>, <__main__
.Article object at 0x046E6090>, <__main__.Article object at 0x046E6130>, <__main__.Article object at 0x046E6170>, <__main__.Article
object at 0x046E6190>, <__main__.Article object at 0x046E61B0>, <__main__.Article object at 0x046E61D0>, <__main__.Article object at
0x046E6150>, <__main__.Article object at 0x046E61F0>, <__main__.Article object at 0x046E6210>, <__main__.Article object at
0x046E6230>, <__main__.Article object at 0x046E6250>, <__main__.Article object at 0x046E6270>, <__main__.Article object at
0x046E6290>, <__main__.Article object at 0x046E62B0>, <__main__.Article object at 0x046E62D0>, <__main__.Article object at
0x046E62F0>, <__main__.Article object at 0x046E6310>, <__main__.Article object at 0x046E6330>, <__main__.Article object at
0x046E6350>, <__main__.Article object at 0x046E6370>, <__main__.Article object at 0x046E6390>, <__main__.Article object at
0x046E63B0>, <__main__.Article object at 0x046E63D0>, <__main__.Article object at 0x046E63F0>, <__main__.Article object at
0x046E6410>, <__main__.Article object at 0x046E6430>, <__main__.Article object at 0x046A5610>]
======================
```

图 3-106　PyCharm 中的运行结果

6. 分组查询

通过 group_by() 方法可以将查询到的结果按照某个字段或多个字段进行分组。下面通过一个示例演示一下如何进行分组查询。

(1) 创建主程序 index.py，代码如下：

```python
# 资源包\Code\chapter3\3.9\20\index.py
from sqlalchemy import create_engine, Column, Integer, String, Enum, func
from sqlalchemy_utils import create_database, database_exists
from sqlalchemy.orm import declarative_base, sessionmaker
HOST = "127.0.0.1"
POST = "3306"
DATABASENAME = "flask_db"
USERNAME = "root"
PASSWORD = "12345678"
DB_URL = f"mysql+pymysql://{USERNAME}:{PASSWORD}@{HOST}:{POST}/{DATABASENAME}"
engine = create_engine(url = DB_URL)
if not database_exists(DB_URL):
```

```python
    create_database(DB_URL)
Base = declarative_base()
class User(Base):
    __tablename__ = "user"
    id = Column(Integer, primary_key = True, autoincrement = True)
    username = Column(String(50), nullable = False)
    age = Column(Integer, default = 0)
    gender = Column(Enum("male", "female", "secret"))
MetaData = Base.metadata
MetaData.drop_all(bind = engine)
MetaData.create_all(bind = engine)
Session = sessionmaker(bind = engine)
db_session = Session()
user1 = User(username = "zhangsan", age = 17, gender = "male")
user2 = User(username = "lisi", age = 18, gender = "male")
user3 = User(username = "wangwu", age = 18, gender = "female")
user4 = User(username = "zhaoliu", age = 19, gender = "female")
user5 = User(username = "sunqi", age = 20, gender = "male")
user6 = User(username = "zhouba", age = 20, gender = "female")
user7 = User(username = "wujiu", age = 21, gender = "male")
user8 = User(username = "zhengshi", age = 21, gender = "female")
db_session.add_all([user1, user2, user3, user4, user5, user6, user7, user8])
db_session.commit()
# 查找每个年龄段对应的人数
results = db_session.query(User.age, func.count(User.id)).group_by(User.age).all()
print(results)
```

[(17, 1), (18, 2), (19, 1), (20, 2), (21, 2)]

图 3-107　PyCharm 中的运行结果

（2）运行上述程序，PyCharm 中的运行结果如图 3-107 所示。

7. 精准查询

数据在进行分组后，通过 having() 方法可以进一步进行精准查询。

下面通过一个示例演示一下如何进行精准查询。

（1）创建主程序 index.py，代码如下：

```python
# 资源包\Code\chapter3\3.9\21\index.py
from sqlalchemy import create_engine, Column, Integer, String, Enum, func
from sqlalchemy_utils import create_database, database_exists
from sqlalchemy.orm import declarative_base, sessionmaker
HOST = "127.0.0.1"
POST = "3306"
DATABASENAME = "flask_db"
USERNAME = "root"
PASSWORD = "12345678"
DB_URL = f"mysql+pymysql://{USERNAME}:{PASSWORD}@{HOST}:{POST}/{DATABASENAME}"
engine = create_engine(url = DB_URL)
if not database_exists(DB_URL):
    create_database(DB_URL)
Base = declarative_base()
class User(Base):
    __tablename__ = "user"
    id = Column(Integer, primary_key = True, autoincrement = True)
    username = Column(String(50), nullable = False)
    age = Column(Integer, default = 0)
```

```
        gender = Column(Enum("male", "female", "secret"))
MetaData = Base.metadata
MetaData.drop_all(bind = engine)
MetaData.create_all(bind = engine)
Session = sessionmaker(bind = engine)
db_session = Session()
user1 = User(username = "zhangsan", age = 17, gender = "male")
user2 = User(username = "lisi", age = 18, gender = "male")
user3 = User(username = "wangwu", age = 18, gender = "female")
user4 = User(username = "zhaoliu", age = 19, gender = "female")
user5 = User(username = "sunqi", age = 20, gender = "male")
user6 = User(username = "zhouba", age = 20, gender = "female")
user7 = User(username = "wujiu", age = 21, gender = "male")
user8 = User(username = "zhengshi", age = 21, gender = "female")
db_session.add_all([user1, user2, user3, user4, user5, user6, user7, user8])
db_session.commit()
#查找年龄小于18岁的人数
results = db_session.query(User.age, func.count(User.id)).group_by(User.age).having(User.
age < 18).all()
print(results)
```

（2）运行上述程序，PyCharm 中的运行结果如图 3-108 所示。

`[(17, 1)]`

8. 多表连接查询

图 3-108　PyCharm 中的运行结果

通过 join()方法可以实现多表连接查询，其语法格式如下：

```
join(right, isouter, full)
```

其中，参数 right 表示连接的右侧数据表，参数 isouter 用于实现左外连接，参数 full 用于实现外连接。

下面通过一个示例演示一下如何进行多表连接查询。

（1）创建主程序 index.py，代码如下：

```
#资源包\Code\chapter3\3.9\22\index.py
from sqlalchemy import create_engine, Column, Integer, String, ForeignKey
from sqlalchemy_utils import create_database, database_exists
from sqlalchemy.orm import declarative_base, sessionmaker, relationship, backref
HOST = "127.0.0.1"
POST = "3306"
DATABASENAME = "flask_db"
USERNAME = "root"
PASSWORD = "12345678"
DB_URL = f"mysql+pymysql://{USERNAME}:{PASSWORD}@{HOST}:{POST}/{DATABASENAME}"
engine = create_engine(url = DB_URL)
if not database_exists(DB_URL):
    create_database(DB_URL)
Base = declarative_base()
class User(Base):
    __tablename__ = "user"
    id = Column(Integer, primary_key = True, autoincrement = True)
    username = Column(String(50), nullable = False)
    extend = relationship("UserExtend", backref = backref('userextend'))
    def __repr__(self):
        return f"< User(id:{self.id},username:{self.username})>"
```

```python
class UserExtend(Base):
    __tablename__ = 'user_extend'
    id = Column(Integer, primary_key = True, autoincrement = True)
    job = Column(String(50))
    uid = Column(Integer, ForeignKey("user.id"))
    def __repr__(self):
        return f"< UserExtend(job:{self.job})>"
MetaData = Base.metadata
MetaData.drop_all(bind = engine)
MetaData.create_all(bind = engine)
Session = sessionmaker(bind = engine)
db_session = Session()
user1 = User(username = '夏正东')
user2 = User(username = '夏兴桂')
user3 = User(username = '于萍')
use_extend1 = UserExtend(job = 'Teacher')
use_extend2 = UserExtend(job = 'civil servant')
db_session.add_all([user1, user2, user3])
db_session.add_all([use_extend1, use_extend2])
db_session.commit()
# 内连接
result = db_session.query(User, UserExtend).join(UserExtend, User.id == UserExtend.id).all()
print(result)
print(' ================ ')
# 左外连接
result1 = db_session.query(User, UserExtend).join(UserExtend, User.id == UserExtend.id, isouter = True).all()
print(result1)
```

（2）运行上述程序，PyCharm 中的运行结果如图 3-109 所示。

```
[[(<User(id:1,username:夏正东)>, <UserExtend(job:Teacher)>), (<User(id:2,username:夏兴桂)>, <UserExtend(job:civil servant)>)]
================
[[(<User(id:1,username:夏正东)>, <UserExtend(job:Teacher)>), (<User(id:2,username:夏兴桂)>, <UserExtend(job:civil servant)>), (<User(id:3,username:于萍)>, None)]
```

图 3-109　PyCharm 中的运行结果

9．子查询

通过 subquery() 方法可以将一个查询对象转换为子查询。

在子查询中，可以通过 label() 方法给相关的字段起别名，而在父查询中，则可以使用这些别名进行条件过滤，并且当需要使用子查询的字段时，可以通过子查询的属性 c 进行获取。

下面通过一个示例演示一下如何进行子查询。

（1）创建主程序 index.py，代码如下：

```python
# 资源包\Code\chapter3\3.9\23\index.py
from sqlalchemy import create_engine, Column, Integer, String
from sqlalchemy_utils import create_database, database_exists
from sqlalchemy.orm import declarative_base, sessionmaker
HOST = "127.0.0.1"
POST = "3306"
DATABASENAME = "flask_db"
```

```python
USERNAME = "root"
PASSWORD = "12345678"
DB_URL = f"mysql+pymysql://{USERNAME}:{PASSWORD}@{HOST}:{POST}/{DATABASENAME}"
engine = create_engine(url=DB_URL)
if not database_exists(DB_URL):
    create_database(DB_URL)
Base = declarative_base()
class User(Base):
    __tablename__ = "user"
    id = Column(Integer, primary_key=True, autoincrement=True)
    username = Column(String(50), nullable=False)
    age = Column(Integer, default=0)
    teach = Column(String(50), nullable=False)
MetaData = Base.metadata
MetaData.drop_all(bind=engine)
MetaData.create_all(bind=engine)
Session = sessionmaker(bind=engine)
db_session = Session()
user1 = User(username="夏正东", age=35, teach='Python')
user2 = User(username="夏兴桂", age=68, teach='Java')
user3 = User(username="于萍", age=68, teach='Linux')
user4 = User(username="侯晓茹", age=35, teach='Python')
db_session.add_all([user1, user2, user3, user4])
db_session.commit()
#寻找与夏正东老师所教的技术一致,并且年龄相同的老师
subuser = db_session.query(User.teach.label("teach"), User.age.label("age")).filter(
    User.username == '夏正东').subquery()
results = db_session.query(User).filter(User.teach == subuser.c.teach, User.age == subuser.c.age).all()
for result in results:
    print(result.username)
```

(2) 运行上述程序,PyCharm 中的运行结果如图 3-110 所示。

```
夏正东
侯晓茹
```

图 3-110　PyCharm 中的运行结果

3.10　Flask-SQLAlchemy

SQLAlchemy 是一个关系数据库框架,提供了 ORM 和原生数据库的相关操作,而 Flask-SQLAlchemy 则是一个对 SQLAlchemy 相关操作进行简化的 Flask 扩展。

3.10.1　安装 Flask-SQLAlchemy

由于 Flask-SQLAlchemy 属于 Python 的第三方库,所以需要进行安装,只需在命令提示符中输入命令 pip install flask-sqlalchemy。

3.10.2　配置 Flask-SQLAlchemy

Flask-SQLAlchemy 的相关配置均被封装到了 Flask 的配置项 app.config 之中,其常

用的配置变量如表 3-11 所示。

表 3-11　常用的配置变量

配置变量	描　　述
SQLALCHEMY_DATABASE_URI	设置数据库的连接地址,格式为协议名://用户名:密码@数据库 IP:端口号/数据库名称
SQLALCHEMY_TRACK_MODIFICATIONS	追踪数据库的变化
SQLALCHEMY_ECHO	打印底层执行的 SQL 语句
SQLALCHEMY_BINDS	当访问多个数据库时,用于设置数据库的连接地址

3.10.3　连接数据库

在 Flask-SQLAlchemy 中连接数据库,需要自定义数据库连接字符串,并将其赋值给对应的配置变量。

下面通过一个示例演示一下如何连接数据库。

(1) 创建名为 flaskProject 的 Flask 项目。

(2) 打开保存在项目根目录下的文件 app.py,代码如下:

```
# 资源包\Code\chapter3\3.10\1\flaskProject\app.py
from flask import Flask
app = Flask(__name__)
HOSTNAME = "127.0.0.1"
PORT = "3306"
DATABASE = "flask_db"
USERNAME = "root"
PASSWORD = "12345678"
DB_URI = f"mysql+pymysql://{USERNAME}:{PASSWORD}@{HOSTNAME}:{PORT}/{DATABASE}"
app.config["SQLALCHEMY_DATABASE_URI"] = DB_URI
app.config["SQLALCHEMY_TRACK_MODIFICATIONS"] = False
@app.route('/')
def index():
    return '连接数据库成功!'
if __name__ == '__main__':
    app.run(debug=True)
```

(3) 运行上述程序,打开浏览器并在网址栏输入 http://127.0.0.1:5000/,其显示内容如图 3-111 所示。

图 3-111　浏览器中的显示内容

3.10.4　获取数据库对象

数据库对象是 Flask-SQLAlchemy 中的一个非常重要的核心对象。SQLAlchemy 中的创建数据库、创建数据表、创建数据表中的字段、CRUD 操作,以及多表间关系等相关操作都是通过其内部封装的相关函数实现的,而在 Flask-SQLAlchemy 中,这些操作则是通过数据库对象的相关属性和方法实现的。

在 Flask-SQLAlchemy 中,通过 flask_sqlalchemy 模块中的 SQLAlchemy 类创建数据库对象。

下面通过一个示例演示一下如何获取数据库对象。

(1) 创建名为 flaskProject 的 Flask 项目。

(2) 打开保存在项目根目录下的文件 app.py,代码如下:

```
# 资源包\Code\chapter3\3.10\2\flaskProject\app.py
from flask import Flask
from flask_sqlalchemy import SQLAlchemy
app = Flask(__name__)
HOSTNAME = "127.0.0.1"
PORT = "3306"
DATABASE = "flask_db"
USERNAME = "root"
PASSWORD = "12345678"
DB_URI = f"mysql+pymysql://{USERNAME}:{PASSWORD}@{HOSTNAME}:{PORT}/{DATABASE}"
app.config["SQLALCHEMY_DATABASE_URI"] = DB_URI
app.config["SQLALCHEMY_TRACK_MODIFICATIONS"] = False
db = SQLAlchemy(app)
@app.route('/')
def index():
    return '获取数据库对象成功'
if __name__ == '__main__':
    app.run(debug = True)
```

(3) 运行上述程序,打开浏览器并在网址栏输入 http://127.0.0.1:5000/,其显示内容如图 3-112 所示。

图 3-112 浏览器中的显示内容

3.10.5 创建数据表

在 Flask-SQLAlchemy 中,用于表示数据表的数据模型必须继承自数据库对象的 Model()方法所创建的类,并且数据模型中的字段需要通过数据库对象的 Column()方法创建,而其内部的字段类型也需要通过数据库对象的相关属性或方法设置。

在完成上述操作后,通过在应用程序上下文(该内容将在后续章节中详细讲解)中使用数据库对象的 create_all()方法将数据模型映射到数据库中。

下面通过一个示例演示一下如何创建数据表。

(1) 创建名为 flaskProject 的 Flask 项目。

(2) 打开保存在项目根目录下的文件 app.py,代码如下:

```python
# 资源包\Code\chapter3\3.10\3\flaskProject\app.py
from flask import Flask
from flask_sqlalchemy import SQLAlchemy
app = Flask(__name__)
HOSTNAME = "127.0.0.1"
PORT = "3306"
DATABASE = "flask_db"
USERNAME = "root"
PASSWORD = "12345678"
DB_URI = f"mysql+pymysql://{USERNAME}:{PASSWORD}@{HOSTNAME}:{PORT}/{DATABASE}"
app.config["SQLALCHEMY_DATABASE_URI"] = DB_URI
app.config["SQLALCHEMY_TRACK_MODIFICATIONS"] = False
db = SQLAlchemy(app)
class User(db.Model):
    __tablename__ = "user"
    id = db.Column(db.Integer, primary_key=True, autoincrement=True)
    username = db.Column(db.String(50), nullable=False)
class Article(db.Model):
    __tablename__ = "article"
    id = db.Column(db.Integer, primary_key=True, autoincrement=True)
    title = db.Column(db.String(50), nullable=False)
    uid = db.Column(db.Integer, db.ForeignKey("user.id"))
with app.app_context():
    db.create_all()
@app.route('/')
def index():
    return '创建数据表成功'
if __name__ == '__main__':
    app.run(debug=True)
```

（3）运行上述程序，打开浏览器并在网址栏输入http://127.0.0.1:5000/，其显示内容如图 3-113 所示。

图 3-113　浏览器中的显示内容

（4）打开命令提示符窗口，登入 MySQL。

（5）在 MySQL 命令行窗口中输入 SQL 语句"desc user;"，其结果如图 3-114 所示。

（6）输入 SQL 语句"desc article;"，其结果如图 3-115 所示。

图 3-114　查看表结构

图 3-115　查看表结构

3.10.6　CRUD 操作

在 Flask-SQLAlchemy 中，可以通过数据库对象的 session() 方法创建会话对象，并通过会话对象的相关方法完成数据的插入、查询、更新和删除等操作。

在完成上述操作后，可以使用会话对象的 commit() 方法将会话中的所有操作作为一个事务提交至数据库。

1. 插入数据

通过会话对象的 add()方法或 add_all()方法即可完成单条数据或多条数据的添加操作。

下面通过一个示例，演示一下如何插入数据。

（1）创建名为 flaskProject 的 Flask 项目。

（2）打开保存在项目根目录下的文件 app.py，代码如下：

```
# 资源包\Code\chapter3\3.10\4\flaskProject\app.py
from flask import Flask
from flask_sqlalchemy import SQLAlchemy
app = Flask(__name__)
HOSTNAME = "127.0.0.1"
PORT = "3306"
DATABASE = "flask_db"
USERNAME = "root"
PASSWORD = "12345678"
DB_URI = f"mysql + pymysql://{USERNAME}:{PASSWORD}@{HOSTNAME}:{PORT}/{DATABASE}"
app.config["SQLALCHEMY_DATABASE_URI"] = DB_URI
app.config["SQLALCHEMY_TRACK_MODIFICATIONS"] = False
db = SQLAlchemy(app)
class User(db.Model):
    __tablename__ = "user"
    id = db.Column(db.Integer, primary_key = True, autoincrement = True)
    username = db.Column(db.String(50), nullable = False)
class Article(db.Model):
    __tablename__ = "article"
    id = db.Column(db.Integer, primary_key = True, autoincrement = True)
    title = db.Column(db.String(50), nullable = False)
    uid = db.Column(db.Integer, db.ForeignKey("user.id"))
with app.app_context():
    db.create_all()
    user1 = User(username = 'xzd')
    user2 = User(username = 'yp')
    user3 = User(username = 'xxg')
    db.session.add(user1)
    db.session.add_all([user2, user3])
    db.session.commit()
@app.route('/')
def index():
    return '插入数据成功'
if __name__ == '__main__':
    app.run(debug = True)
```

（3）运行上述程序，打开浏览器并在网址栏输入 http://127.0.0.1:5000/，其显示内容如图 3-116 所示。

（4）打开命令提示符窗口，登入 MySQL。

（5）在 MySQL 命令行窗口中输入 SQL 语句"select * from user;"，其结果如图 3-117 所示。

2. 查询数据

首先通过"数据模型.query"的方式获取查询对象，然后通过查询对象的相关方法即可获取查询结果。

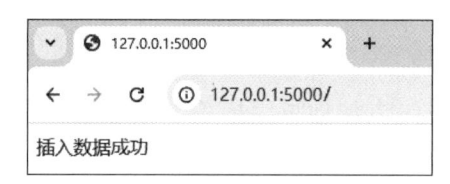

图 3-116　浏览器中的显示内容　　　　图 3-117　查询数据表中的数据

下面通过一个示例演示一下如何查询数据。

(1) 创建名为 flaskProject 的 Flask 项目。

(2) 打开保存在项目根目录下的文件 app.py，代码如下：

```python
# 资源包\Code\chapter3\3.10\5\flaskProject\app.py
from flask import Flask
from flask_sqlalchemy import SQLAlchemy
app = Flask(__name__)
HOSTNAME = "127.0.0.1"
PORT = "3306"
DATABASE = "flask_db"
USERNAME = "root"
PASSWORD = "12345678"
DB_URI = f"mysql+pymysql://{USERNAME}:{PASSWORD}@{HOSTNAME}:{PORT}/{DATABASE}"
app.config["SQLALCHEMY_DATABASE_URI"] = DB_URI
app.config["SQLALCHEMY_TRACK_MODIFICATIONS"] = False
db = SQLAlchemy(app)
class User(db.Model):
    __tablename__ = "user"
    id = db.Column(db.Integer, primary_key=True, autoincrement=True)
    username = db.Column(db.String(50), nullable=False)
class Article(db.Model):
    __tablename__ = "article"
    id = db.Column(db.Integer, primary_key=True, autoincrement=True)
    title = db.Column(db.String(50), nullable=False)
    uid = db.Column(db.Integer, db.ForeignKey("user.id"))
with app.app_context():
    db.create_all()
    user1 = User(username='xzd')
    user2 = User(username='yp')
    user3 = User(username='xxg')
    db.session.add(user1)
    db.session.add_all([user2, user3])
    db.session.commit()
    user_first = User.query.first()
    print(user_first.username)
    print('==========')
    users = User.query.all()
    for user in users:
        print(user.username)
    print('==========')
    user_yp = User.query.filter(User.username == "yp").first()
    print(user_yp.username)
@app.route('/')
def index():
    return '查询数据成功'
if __name__ == '__main__':
    app.run(debug=True)
```

(3)运行上述程序,PyCharm 中的运行结果如图 3-118 所示。

3．删除数据

通过会话对象的 delete() 方法即可完成数据的删除操作。
下面通过一个示例演示一下如何删除数据。

(1)创建名为 flaskProject 的 Flask 项目。

(2)打开保存在项目根目录下的文件 app.py,代码如下:

图 3-118　PyCharm 中的运行结果

```
# 资源包\Code\chapter3\3.10\6\flaskProject\app.py
from flask import Flask
from flask_sqlalchemy import SQLAlchemy
app = Flask(__name__)
HOSTNAME = "127.0.0.1"
PORT = "3306"
DATABASE = "flask_db"
USERNAME = "root"
PASSWORD = "12345678"
DB_URI = f"mysql+pymysql://{USERNAME}:{PASSWORD}@{HOSTNAME}:{PORT}/{DATABASE}"
app.config["SQLALCHEMY_DATABASE_URI"] = DB_URI
app.config["SQLALCHEMY_TRACK_MODIFICATIONS"] = False
db = SQLAlchemy(app)
class User(db.Model):
    __tablename__ = "user"
    id = db.Column(db.Integer, primary_key=True, autoincrement=True)
    username = db.Column(db.String(50), nullable=False)
class Article(db.Model):
    __tablename__ = "article"
    id = db.Column(db.Integer, primary_key=True, autoincrement=True)
    title = db.Column(db.String(50), nullable=False)
    uid = db.Column(db.Integer, db.ForeignKey("user.id"))
with app.app_context():
    db.create_all()
    user1 = User(username='xzd')
    user2 = User(username='yp')
    user3 = User(username='xxg')
    db.session.add(user1)
    db.session.add_all([user2, user3])
    db.session.commit()
    user_first = User.query.first()
    db.session.delete(user_first)
    db.session.commit()
@app.route('/')
def index():
    return '删除数据成功'
if __name__ == '__main__':
    app.run(debug=True)
```

图 3-119　查询数据表中的数据

(3)运行上述程序。

(4)打开命令提示符窗口,登入 MySQL。

(5)在 MySQL 命令行窗口中输入 SQL 语句"select * from user;",其结果如图 3-119 所示。

4．修改数据

对数据库中的数据进行修改,只需修改查询出的数据对象的相关属性。

下面通过一个示例演示一下如何修改数据。

（1）创建名为 flaskProject 的 Flask 项目。

（2）打开保存在项目根目录下的文件 app.py，代码如下：

```python
# 资源包\Code\chapter3\3.10\7\flaskProject\app.py
from flask import Flask
from flask_sqlalchemy import SQLAlchemy
app = Flask(__name__)
HOSTNAME = "127.0.0.1"
PORT = "3306"
DATABASE = "flask_db"
USERNAME = "root"
PASSWORD = "12345678"
DB_URI = f"mysql+pymysql://{USERNAME}:{PASSWORD}@{HOSTNAME}:{PORT}/{DATABASE}"
app.config["SQLALCHEMY_DATABASE_URI"] = DB_URI
app.config["SQLALCHEMY_TRACK_MODIFICATIONS"] = False
db = SQLAlchemy(app)
class User(db.Model):
    __tablename__ = "user"
    id = db.Column(db.Integer, primary_key=True, autoincrement=True)
    username = db.Column(db.String(50), nullable=False)
class Article(db.Model):
    __tablename__ = "article"
    id = db.Column(db.Integer, primary_key=True, autoincrement=True)
    title = db.Column(db.String(50), nullable=False)
    uid = db.Column(db.Integer, db.ForeignKey("user.id"))
# 插入数据
# with app.app_context():
    # db.create_all()
    # user1 = User(username='xzd')
    # user2 = User(username='yp')
    # user3 = User(username='xxg')
    # db.session.add(user1)
    # db.session.add_all([user2, user3])
    # db.session.commit()
@app.route('/')
def index():
    user_first = User.query.first()
    print(user_first.username)
    print('==================')
    user_first.username = 'mother_yp'
    user_first_new = User.query.first()
    print(user_first_new.username)
    db.session.commit()
    return '数据修改成功'
if __name__ == '__main__':
    app.run(debug=True)
```

```
yp
==================
mother_yp
```

图 3-120　PyCharm 中的运行结果

（3）运行上述程序，PyCharm 中的运行结果如图 3-120 所示。

3.10.7　多表间关系

在 Flask-SQLAlchemy 中，多表间关系通过数据库对象的

relationship()方法实现。

1. 一对多关系

下面通过一个示例演示一下如何实现一对多关系。

（1）创建名为flaskProject的Flask项目。

（2）打开保存在项目根目录下的文件app.py，代码如下：

```python
# 资源包\Code\chapter3\3.10\8\flaskProject\app.py
from flask import Flask
from flask_sqlalchemy import SQLAlchemy
app = Flask(__name__)
HOSTNAME = "127.0.0.1"
PORT = "3306"
DATABASE = "flask_db"
USERNAME = "root"
PASSWORD = "12345678"
DB_URI = f"mysql+pymysql://{USERNAME}:{PASSWORD}@{HOSTNAME}:{PORT}/{DATABASE}"
app.config["SQLALCHEMY_DATABASE_URI"] = DB_URI
app.config["SQLALCHEMY_TRACK_MODIFICATIONS"] = False
db = SQLAlchemy(app)
class User(db.Model):
    __tablename__ = "user"
    id = db.Column(db.Integer, primary_key=True, autoincrement=True)
    username = db.Column(db.String(50), nullable=False)
    articles = db.relationship(argument="Article")
class Article(db.Model):
    __tablename__ = "article"
    id = db.Column(db.Integer, primary_key=True, autoincrement=True)
    title = db.Column(db.String(50), nullable=False)
    content = db.Column(db.Text, nullable=False)
    uid = db.Column(db.Integer, db.ForeignKey("user.id"))
    author = db.relationship("User")
with app.app_context():
    db.create_all()
    user = User(username='夏正东')
    db.session.add(user)
    db.session.commit()
    article1 = Article(title='《Python全栈开发——Web编程》', content='《Python全栈开发——Web编程》', uid=1)
    article2 = Article(title='《Python全栈开发——数据分析》', content='《Python全栈开发——数据分析》', uid=1)
    article3 = Article(title='《Python全栈开发——高阶编程》', content='《Python全栈开发——高阶编程》', uid=1)
    db.session.add_all([article1, article2, article3])
    db.session.commit()
    article = Article.query.first()
    print(article.author.username)
    user = User.query.first()
    for article in user.articles:
        print(article.title)
    db.session.commit()
@app.route('/')
def index():
    return '一对多关系'
if __name__ == '__main__':
    app.run(debug=True)
```

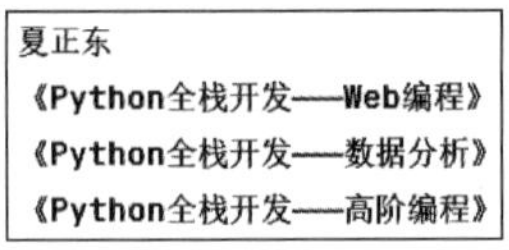

图 3-121 PyCharm 中的运行结果

(3) 运行上述程序，PyCharm 中的运行结果如图 3-121 所示。

2. 一对一关系

下面通过一个示例演示一下如何实现一对一关系。

(1) 创建名为 flaskProject 的 Flask 项目。

(2) 打开保存在项目根目录下的文件 app.py，代码如下：

```python
# 资源包\Code\chapter3\3.10\9\flaskProject\app.py
from flask import Flask
from flask_sqlalchemy import SQLAlchemy
app = Flask(__name__)
HOSTNAME = "127.0.0.1"
PORT = "3306"
DATABASE = "flask_db"
USERNAME = "root"
PASSWORD = "12345678"
DB_URI = f"mysql+pymysql://{USERNAME}:{PASSWORD}@{HOSTNAME}:{PORT}/{DATABASE}"
app.config["SQLALCHEMY_DATABASE_URI"] = DB_URI
app.config["SQLALCHEMY_TRACK_MODIFICATIONS"] = False
db = SQLAlchemy(app)
class User(db.Model):
    __tablename__ = "user"
    id = db.Column(db.Integer, primary_key=True, autoincrement=True)
    username = db.Column(db.String(50), nullable=False)
    extend = db.relationship("UserExtend", uselist=False)
class UserExtend(db.Model):
    __tablename__ = 'user_extend'
    id = db.Column(db.Integer, primary_key=True, autoincrement=True)
    job = db.Column(db.String(50))
    uid = db.Column(db.Integer, db.ForeignKey("user.id"))
    user = db.relationship("User")
    # user = relationship("User", backref=backref("extend", uselist=False))
class Article(db.Model):
    __tablename__ = "article"
    id = db.Column(db.Integer, primary_key=True, autoincrement=True)
    title = db.Column(db.String(50), nullable=False)
    content = db.Column(db.Text, nullable=False)
    uid = db.Column(db.Integer, db.ForeignKey("user.id"))
    author = db.relationship(argument="User", backref="articles")
with app.app_context():
    db.create_all()
    user = User(username='夏正东')
    use_extend = UserExtend(job='Teacher')
    user.extend = use_extend
    db.session.add(user)
    db.session.commit()
@app.route('/')
def index():
    return '一对一关系'
if __name__ == '__main__':
    app.run(debug=True)
```

(3) 运行上述程序。

(4) 打开命令提示符窗口，登入 MySQL。

(5) 在 MySQL 命令行窗口中输入 SQL 语句"select * from user;",其结果如图 3-122 所示。
(6) 输入 SQL 语句"select * from user_extend;",其结果如图 3-123 所示。

图 3-122　查询数据表中的数据　　图 3-123　查询数据表中的数据

3. 多对多关系

下面通过一个示例演示一下如何实现多对多关系。

(1) 创建名为 flaskProject 的 Flask 项目。
(2) 打开保存在项目根目录下的文件 app.py,代码如下:

```python
#资源包\Code\chapter3\3.10\10\flaskProject\app.py
from flask import Flask
from flask_sqlalchemy import SQLAlchemy
app = Flask(__name__)
HOSTNAME = "127.0.0.1"
PORT = "3306"
DATABASE = "flask_db"
USERNAME = "root"
PASSWORD = "12345678"
DB_URI = f"mysql + pymysql://{USERNAME}:{PASSWORD}@{HOSTNAME}:{PORT}/{DATABASE}"
app.config["SQLALCHEMY_DATABASE_URI"] = DB_URI
app.config["SQLALCHEMY_TRACK_MODIFICATIONS"] = False
db = SQLAlchemy(app)
article_tag = db.Table(
    "article_tag",
    db.Column("article_id", db.Integer, db.ForeignKey("article.id"), primary_key = True),
    db.Column("tag_id", db.Integer, db.ForeignKey("tag.id"), primary_key = True)
)
class Article(db.Model):
    __tablename__ = "article"
    id = db.Column(db.Integer, primary_key = True, autoincrement = True)
    title = db.Column(db.String(50), nullable = False)
    tags = db.relationship('Tag', backref = 'articles', secondary = article_tag)
class Tag(db.Model):
    __tablename__ = "tag"
    id = db.Column(db.Integer, primary_key = True, autoincrement = True)
    name = db.Column(db.String(50), nullable = False)
with app.app_context():
    db.create_all()
    article1 = Article(title = "Python全栈开发——数据分析")
    article2 = Article(title = "Python全栈开发——Web编程")
    tag1 = Tag(name = "Python")
    tag2 = Tag(name = "夏正东")
    article1.tags.append(tag1)
    article1.tags.append(tag2)
    article2.tags.append(tag1)
    article2.tags.append(tag2)
    db.session.add(article1)
    db.session.add(article2)
```

```
        db.session.commit()
    article = Article.query.first()
    for tag in article.tags:
        print(tag.name)
    print('==============')
    tag = Tag.query.first()
    for article in tag.articles:
        print(article.title)
@app.route('/')
def index():
    return '多对多关系'
if __name__ == '__main__':
    app.run(debug = True)
```

（3）运行上述程序，PyCharm 中的运行结果如图 3-124 所示。

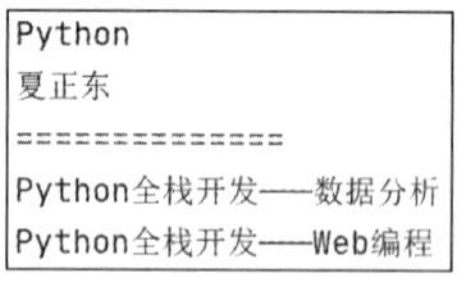

```
Python
夏正东
==============
Python全栈开发——数据分析
Python全栈开发——Web编程
```

图 3-124　PyCharm 中的运行结果

3.11　Alembic

Alembic 是一款轻量型的数据库迁移工具，其使用 SQLAlchemy 作为数据库引擎，为关系型数据提供创建、管理、更改，以及调用等操作的脚本，协助开发和运维人员在系统上线后对数据库进行在线管理。

3.11.1　安装 Alembic

由于 Alembic 属于 Python 的第三方库，所以需要进行安装，只需在命令提示符中输入命令 pip install alembic。

3.11.2　Alembic 操作

1. 初始化 Alembic 仓库

（1）创建名为 flaskProject 的 Flask 项目。

（2）在项目的根目录下创建数据模型文件 model.py，代码如下：

```
# 资源包\Code\chapter3\3.11\1\flaskProject\model.py
from sqlalchemy import create_engine, Column, Integer, String
from sqlalchemy_utils import create_database, database_exists
from sqlalchemy.orm import declarative_base
HOST = "127.0.0.1"
POST = "3306"
DATABASENAME = "flask_db"
USERNAME = "root"
PASSWORD = "12345678"
DB_URL = f"mysql+pymysql://{USERNAME}:{PASSWORD}@{HOST}:{POST}/{DATABASENAME}"
```

```
engine = create_engine(url=DB_URL)
if not database_exists(DB_URL):
    create_database(DB_URL)
Base = declarative_base()
class User(Base):
    __tablename__ = "user"
    id = Column(Integer, primary_key=True, autoincrement=True)
    username = Column(String(50), nullable=False)
MetaData = Base.metadata
```

（3）进入项目所在的目录中，激活虚拟环境，并输入命令"alembic init 仓库名"，即可完成 Alembic 仓库的初始化工作，如图 3-125 所示。

```
E:\Python全栈开发\flaskProject>workon flask_env
(flask_env) E:\Python全栈开发\flaskProject>alembic init alembic
Creating directory 'E:\\Python全栈开发\\flaskProject\\alembic' ... done
Creating directory 'E:\\Python全栈开发\\flaskProject\\alembic\\versions' ... done
Generating E:\Python全栈开发\flaskProject\alembic.ini ... done
Generating E:\Python全栈开发\flaskProject\alembic\env.py ... done
Generating E:\Python全栈开发\flaskProject\alembic\README ... done
Generating E:\Python全栈开发\flaskProject\alembic\script.py.mako ... done
Please edit configuration/connection/logging settings in 'E:\\Python全栈开发\\flaskProject\\alembic.ini' before proceeding.
```

图 3-125　初始化 Alembic 仓库

Alembic 仓库初始化后，项目所在的目录下会生成如图 3-126 所示的目录和文件。

其中，文件夹 alembic 表示 Alembic 仓库的根目录；文件夹 versions 用于存放各版本的迁移脚本，初始情况下为空目录；文件 env.py 包含配置和生成 SQLAlchemy 引擎的指令；文件 README 为信息文件；文件 script.py.mako 用于生成新的迁移脚本文件；文件 alembic.ini 用于定义 Alembic 的运行参数和环境变量。

图 3-126　Alembic 仓库的文件结构

2. 修改 Alembic 配置文件

Alembic 仓库被成功初始化之后，需要对其内部的配置文件进行修改。

1) alembic.ini 文件

在该文件中，主要进行数据库连接的相关配置，即设置 sqlalchemy.url 的值，其格式如下：

```
driver://username:password@localhost/dbname
```

其中，参数 driver 表示数据库驱动，参数 username 表示登录数据库的用户名，参数 password 表示登录数据库的密码，参数 localhost 表示数据库的 URL，参数 dbname 表示待连接的数据库，代码如下：

```
# 资源包\Code\chapter3\3.11\1\flaskProject\alembic.ini
sqlalchemy.url = mysql+pymysql://root:12345678@localhost/flask_db?charset=utf8
```

2) env.py 文件

在该文件中，主要进行变量 target_metadata 值的相关配置，其值必须为 MetaData 对象，代码如下：

```python
#资源包\Code\chapter3\3.11\1\flaskProject\alembic\env.py
import sys, os
sys.path.append(os.path.dirname(os.path.dirname(__file__)))
import model
target_metadata = model.Base.metadata
```

3. 生成迁移脚本文件

在当前虚拟环境下，使用alembic的子命令revision即可生成新的迁移脚本文件。该命令具有两个参数：一是--autogenerate，表示自动将数据模型的修改生成迁移脚本；二是-m，表示迁移脚本的描述，如图3-127所示。

图3-127　生成迁移脚本文件

此时，如果打开Alembic仓库根目录中的versions目录，则会发现新增了一条迁移脚本文件，如图3-128所示。

图3-128　新增的迁移脚本文件

4. 更新数据库

在生成迁移脚本文件后，通过alembic的子命令upgrade即可将指定版本的迁移脚本映射到数据库中，如图3-129所示。

图3-129　更新数据库

这里需要注意的是，该命令会执行指定版本的迁移脚本文件中的upgrade()函数。

此时，查看数据库中所创建的表结构，如图3-130所示。

图3-130　查看表结构

5. 修改数据模型

当数据模型中的内容发生变动时，需要重新执行生成迁移脚本文件和更新数据库。

下面通过一个示例演示一下如何在数据模型中新增字段。

（1）更改上一个保存在项目根目录下的文件model.py，为数据模型新增字段，代码如下：

```python
#资源包\Code\chapter3\3.11\2\flaskProject\model.py
from sqlalchemy import create_engine, Column, Integer, String
from sqlalchemy_utils import create_database, database_exists
from sqlalchemy.orm import declarative_base
HOST = "127.0.0.1"
POST = "3306"
DATABASENAME = "flask_db"
```

```
USERNAME = "root"
PASSWORD = "12345678"
DB_URL = f"mysql+pymysql://{USERNAME}:{PASSWORD}@{HOST}:{POST}/{DATABASENAME}"
engine = create_engine(url=DB_URL)
if not database_exists(DB_URL):
    create_database(DB_URL)
Base = declarative_base()
class User(Base):
    __tablename__ = "user"
    id = Column(Integer, primary_key=True, autoincrement=True)
    username = Column(String(50), nullable=False)
    #新增字段
    age = Column(Integer, default=0)
MetaData = Base.metadata
```

（2）打开命令提示符窗口，进入项目所在的目录中，激活虚拟环境，并输入 Alembic 的子命令 revision 生成迁移脚本文件，如图 3-131 所示。

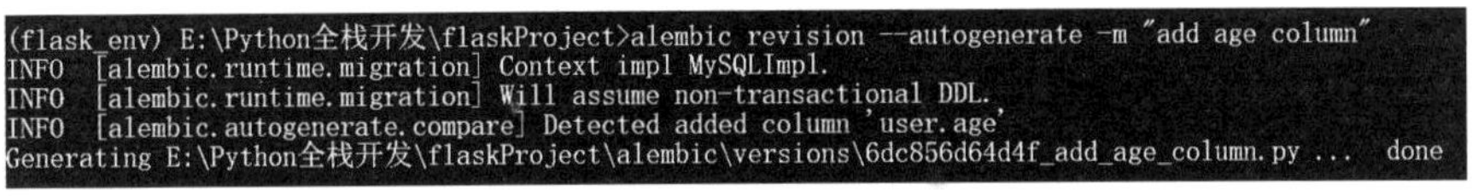

```
(flask_env) E:\Python全栈开发\flaskProject>alembic revision --autogenerate -m "add age column"
INFO  [alembic.runtime.migration] Context impl MySQLImpl.
INFO  [alembic.runtime.migration] Will assume non-transactional DDL.
INFO  [alembic.autogenerate.compare] Detected added column 'user.age'
Generating E:\Python全栈开发\flaskProject\alembic\versions\6dc856d64d4f_add_age_column.py ...  done
```

图 3-131 生成迁移脚本文件

（3）使用 Alembic 的子命令 upgrade 更新数据库，如图 3-132 所示。

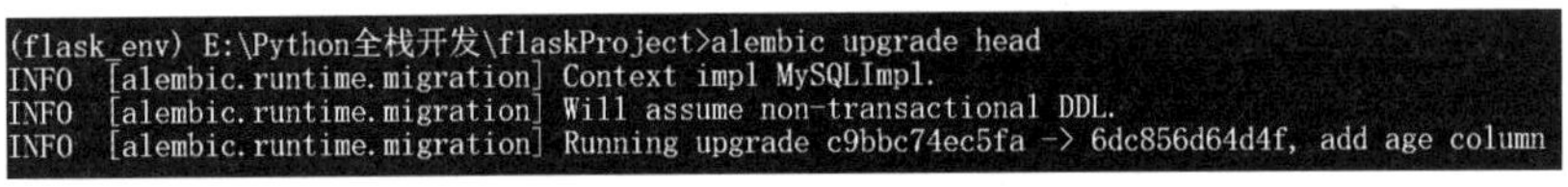

```
(flask_env) E:\Python全栈开发\flaskProject>alembic upgrade head
INFO  [alembic.runtime.migration] Context impl MySQLImpl.
INFO  [alembic.runtime.migration] Will assume non-transactional DDL.
INFO  [alembic.runtime.migration] Running upgrade c9bbc74ec5fa -> 6dc856d64d4f, add age column
```

图 3-132 更新数据库

（4）打开命令提示符窗口，登入 MySQL，并输入 SQL 语句“desc user;”，查看 user 表的结构，如图 3-133 所示。

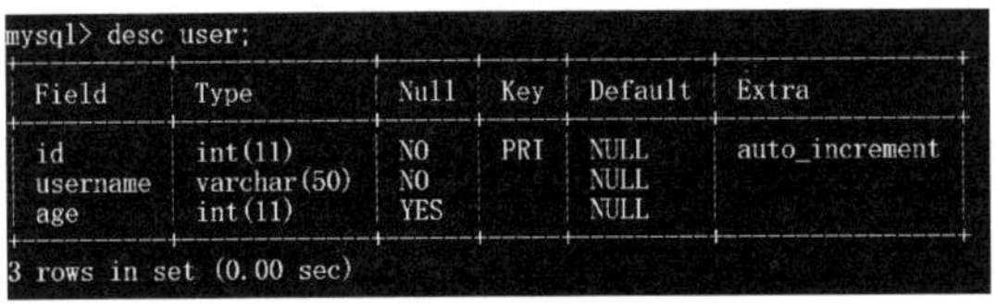

```
mysql> desc user;
+----------+-------------+------+-----+---------+----------------+
| Field    | Type        | Null | Key | Default | Extra          |
+----------+-------------+------+-----+---------+----------------+
| id       | int(11)     | NO   | PRI | NULL    | auto_increment |
| username | varchar(50) | NO   |     | NULL    |                |
| age      | int(11)     | YES  |     | NULL    |                |
+----------+-------------+------+-----+---------+----------------+
3 rows in set (0.00 sec)
```

图 3-133 查看表结构

6. Alembic 常用的子命令

除了上述讲解的 Alembic 子命令，在表 3-12 中还展示了 Alembic 的其他常用子命令。

表 3-12 Alembic 常用的子命令

命　令	描　述
downgrade	将数据库降级至指定版本，其会执行指定版本的迁移脚本中的 downgrade()函数
head	当前迁移脚本的版本号
heads	展示 head 所指向的迁移脚本的版本号
history	展示所有的迁移脚本的版本号及其信息
current	展示当前数据库中的迁移脚本的版本号

3.11.3　在 Flask-SQLAlchemy 中操作 Alembic

1. 初始化 Alembic 仓库

（1）创建名为 flaskProject 的 Flask 项目。

（2）打开保存在项目根目录下的文件 app.py，代码如下：

```python
# 资源包\Code\chapter3\3.11\3\flaskProject\app.py
from flask import Flask
from flask_sqlalchemy import SQLAlchemy
app = Flask(__name__)
HOSTNAME = "127.0.0.1"
PORT = "3306"
DATABASE = "flask_db"
USERNAME = "root"
PASSWORD = "12345678"
DB_URI = f"mysql+pymysql://{USERNAME}:{PASSWORD}@{HOSTNAME}:{PORT}/{DATABASE}"
app.config["SQLALCHEMY_DATABASE_URI"] = DB_URI
app.config["SQLALCHEMY_TRACK_MODIFICATIONS"] = False
db = SQLAlchemy(app)
class User(db.Model):
    __tablename__ = "user"
    id = db.Column(db.Integer, primary_key=True, autoincrement=True)
    username = db.Column(db.String(50), nullable=False)
@app.route('/')
def hello_world():
    return '老夏学院'
if __name__ == '__main__':
    app.run(debug=True)
```

（3）打开命令提示符窗口，进入项目所在的目录中，激活虚拟环境，并输入命令"alembic init 仓库名"，即可完成 Alembic 仓库的初始化工作，如图 3-134 所示。

```
E:\Python全栈开发\flaskProject>workon flask_env
(flask_env) E:\Python全栈开发\flaskProject>alembic init alembic
Creating directory 'E:\\Python全栈开发\\flaskProject\\alembic' ...  done
Creating directory 'E:\\Python全栈开发\\flaskProject\\alembic\\versions' ...  done
Generating E:\Python全栈开发\flaskProject\alembic.ini ...  done
Generating E:\Python全栈开发\flaskProject\alembic\env.py ...  done
Generating E:\Python全栈开发\flaskProject\alembic\README ...  done
Generating E:\Python全栈开发\flaskProject\alembic\script.py.mako ...  done
Please edit configuration/connection/logging settings in 'E:\\Python全栈开发\\flaskProject\\alembic.ini' before proceeding.
```

图 3-134　初始化 Alembic 仓库

Alembic 仓库初始化后，项目所在的目录下会生成如图 3-135 所示的目录和文件。

图 3-135　Alembic 仓库的文件结构

2. 修改 Alembic 配置文件

1）alembic.ini 文件

在该文件中，主要进行数据库连接的相关配置，即设置 sqlalchemy.url 的值，其格式如下：

```
driver://username:password@localhost/dbname
```

其中，参数 driver 表示数据库驱动，参数 username 表示登录数据库的用户名，参数 password 表示登录数据库的密码，参数 localhost 表示数据库的 URL，参数 dbname 表示待连接的数据库，代码如下：

```
# 资源包\Code\chapter3\3.11\3\flaskProject\alembic.ini
sqlalchemy.url = mysql+pymysql://root:12345678@localhost/flask_db?charset=utf8
```

2) env.py 文件

在该文件中，主要进行变量 target_metadata 值的相关配置，其值必须为 MetaData 对象，代码如下：

```
# 资源包\Code\chapter3\3.11\3\flaskProject\alembic\env.py
import sys, os
sys.path.append(os.path.dirname(os.path.dirname(__file__)))
import app
target_metadata = app.db.Model.metadata
```

3. 生成迁移脚本文件

在当前虚拟环境下，使用 Alembic 的子命令 revision 即可生成新的迁移脚本文件。该命令具有两个参数，一是--autogenerate，表示自动将模型的修改生成迁移脚本；二是-m，表示迁移脚本的描述，如图 3-136 所示。

图 3-136　生成迁移脚本文件

此时，如果打开 Alembic 仓库根目录中的 versions 目录，则会发现新增了一条迁移脚本文件，如图 3-137 所示。

图 3-137　新增的迁移脚本文件

4. 更新数据库

在生成迁移脚本文件后，通过 alembic 的子命令 upgrade 即可将指定版本的迁移脚本映射到数据库中，如图 3-138 所示。

图 3-138　更新数据库

需要注意的是，该命令会执行指定版本的迁移脚本文件中的 upgrade() 函数。

此时，查看数据库中所创建的表结构，如图 3-139 所示。

图 3-139　查看表结构

3.12　Flask-Script

虽然 Flask 的 Web 服务器支持多种启动设置选项，但这些选项都只能在脚本中作为参数传递给 run() 函数，而传递设置选项最理想的方式是使用命令行解析器，Flask-Scrip 就是这么一个 Flask 扩展，为 Flask 程序添加一个命令行解析器。

Flask-Script 主要用于生成 Shell 命令，为在 Flask 里编写额外的脚本提供支持，包括运行 Web 服务器、定制 Python 命令行、执行数据库初始化、定时任务或其他运行在 Web 应用之外的命令行任务的脚本。

3.12.1　安装 Flask-Script

由于 Flask-Script 属于 Python 的第三方库，所以需要进行安装，只需在命令提示符中输入命令 pip install flask-script。

3.12.2　创建自定义命令

自定义命令通过 Manager 类的实例对象的装饰器或相关方法实现。

1. 使用装饰器@command

通过 Manager 类的实例对象的装饰器@command 的方式可以快速地创建自定义命令。

下面通过一个示例演示一下如何创建自定义命令。

（1）创建名为 flaskProject 的 Flask 项目。

（2）在项目的根目录下创建文件 manage.py，代码如下：

```python
# 资源包\Code\chapter3\3.12\1\flaskProject\manage.py
from flask_script import Manager
from app import app
# 使用 Manager 类创建对象
manager = Manager(app)
@manager.command
def greet():
    print("欢迎光临老夏学院")
if __name__ == '__main__':
    manager.run()
```

（3）打开命令提示符窗口，进入项目根目录下，激活虚拟环境，并输入命令 python manage.py greet，即可获取命令 greet 所对应的内容，如图 3-140 所示。

```
(flask_env) E:\Python全栈开发\flaskProject>python manage.py greet
欢迎光临老夏学院
```

图 3-140　命令 greet 所对应的内容

2. 使用装饰器@option

通过 Manager 类的实例对象的装饰器@option 的方式可以快速地创建带参数的自定义命令。

下面通过一个示例演示一下如何创建带参数的自定义命令。

（1）创建名为 flaskProject 的 Flask 项目。

（2）打开保存在项目根目录下的文件 app.py，代码如下：

```python
# 资源包\Code\chapter3\3.12\2\flaskProject\manage.py
from flask import Flask
```

```python
from flask_sqlalchemy import SQLAlchemy
app = Flask(__name__)
HOSTNAME = "127.0.0.1"
PORT = "3306"
DATABASE = "flask_db"
USERNAME = "root"
PASSWORD = "12345678"
DB_URI = f"mysql+pymysql://{USERNAME}:{PASSWORD}@{HOSTNAME}:{PORT}/{DATABASE}"
app.config["SQLALCHEMY_DATABASE_URI"] = DB_URI
app.config["SQLALCHEMY_TRACK_MODIFICATIONS"] = False
db = SQLAlchemy(app)
class BackendUser(db.Model):
    __tablename__ = "backend_user"
    id = db.Column(db.Integer, primary_key=True, autoincrement=True)
    username = db.Column(db.String(50), nullable=False)
    email = db.Column(db.String(50), nullable=False)
with app.app_context():
    db.create_all()
@app.route('/')
def hello_world():
    return '老夏学院'
if __name__ == '__main__':
    app.run(debug=True)
```

（3）在项目的根目录下创建文件 manage.py，代码如下：

```python
# 资源包\Code\chapter3\3.12\2\flaskProject\manage.py
from flask_script import Manager
from app import app, BackendUser, db
manager = Manager(app)
@manager.option("-u", "--username", dest="username")
@manager.option("-e", "--email", dest="email")
def add_user(username, email):
    user = BackendUser(username=username, email=email)
    db.session.add(user)
    db.session.commit()
if __name__ == '__main__':
    manager.run()
```

（4）打开命令提示符窗口，进入项目根目录下，激活虚拟环境，并输入命令"python manage.py add_user -u xzd -e xiazhengdong@vip.qq.com"，即可将数据插入数据库中。

（5）打开命令提示符窗口，登入 MySQL。

（6）在 MySQL 命令行窗口中输入 SQL 语句"select * from backend_user;"，其结果如图 3-141 所示。

图 3-141　查看数据库中的数据

3. 定义 Command 子类

通过 Manager 类的实例对象的 add_command() 方法可以将 Command 子类映射为自定义的命令。

下面通过一个示例演示一下如何定义 Command 子类。

（1）创建名为 flaskProject 的 Flask 项目。

（2）在项目的根目录下创建文件 manage.py，代码如下：

```
# 资源包\Code\chapter3\3.12\3\flaskProject\manage.py
from flask_script import Manager, Command
from app import app
manager = Manager(app)
class Greet(Command):
    def run(self):
        print("欢迎光临老夏学院")
manager.add_command('greet', Greet())
if __name__ == '__main__':
    manager.run()
```

(3) 打开命令提示符窗口,进入项目根目录下,激活虚拟环境,并输入命令 python manage.py greet,即可获取命令 greet 所对应的内容,如图 3-142 所示。

```
(flask_env) E:\Python全栈开发\flaskProject>python manage.py greet
欢迎光临老夏学院
```

图 3-142　命令 greet 所对应的内容

此外,通过 add_command()方法还可以创建子命令。

下面通过一个示例演示一下如何创建子命令。

(1) 创建名为 flaskProject 的 Flask 项目。

(2) 在项目的根目录下创建文件 manage.py,代码如下:

```
# 资源包\Code\chapter3\3.12\4\flaskProject\manage.py
from flask_script import Manager
from app import app
from db_script import db_manager
manager = Manager(app)
manager.add_command('db', db_manager)
if __name__ == '__main__':
    manager.run()
```

(3) 在项目的根目录下创建文件 db_script.py,代码如下:

```
# 资源包\Code\chapter3\3.12\4\flaskProject\db_script.py
from flask_script import Manager
db_manager = Manager()
@db_manager.command
def init():
    print('初始化仓库-finished')
@db_manager.command
def revision():
    print("生成迁移脚本-finished")
@db_manager.command
def upgrade():
    print("数据库更新-finished")
```

(4) 打开命令提示符窗口,进入项目根目录下,激活虚拟环境,并输入命令 python manage.py db init,即可获取子命令 init 所对应的内容,如图 3-143 所示。

```
(flask_env) E:\Python全栈开发\flaskProject>python manage.py db init
初始化仓库-finished
```

图 3-143　子命令 init 所对应的内容

3.13 Flask-Migrate

Flask-Migrate 是基于 Alembic 的封装，并集成到 Flask 中，用于处理数据库迁移与映射。

3.13.1 安装 Flask-Migrate

由于 Flask-Migrate 属于 Python 的第三方库，所以需要进行安装，只需在命令提示符中输入命令 pip install flask-migrate。

3.13.2 Flask-Migrate 操作

在进行 Flask-Migrate 的相关操作前，需要对项目的结构进行重构，如图 3-144 所示。其中，文件 app.py 为项目的主程序，文件 models.py 用于编写数据模型，文件 exts.py 用于定义数据库对象，以防止循环引用的发生，文件 config.py 用于编写数据库连接的相关参数，文件 manage.py 用于编写数据库迁移与映射的相关命令。

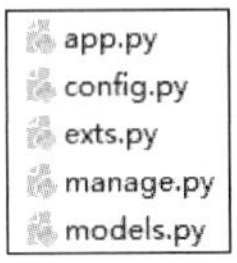

图 3-144 项目结构重构

下面通过一个示例演示一下如何使用 Flask-Migrate 操作数据库。

(1) 创建名为 flaskProject 的 Flask 项目。

(2) 打开保存在项目根目录下的文件 app.py，代码如下：

```
# 资源包\Code\chapter3\3.13\1\flaskProject\app.py
from flask import Flask
import config
from exts import db
app = Flask(__name__)
app.config.from_object(config)
# 初始化数据库对象
db.init_app(app)
@app.route('/')
def hello_world():
    return '老夏学院'
if __name__ == '__main__':
    app.run(debug = True)
```

(3) 在项目的根目录下创建文件 config.py，代码如下：

```
# 资源包\Code\chapter3\3.13\1\flaskProject\config.py
HOSTNAME = "127.0.0.1"
PORT = "3306"
DATABASE = "flask_db"
USERNAME = "root"
PASSWORD = "12345678"
DB_URI = f"mysql+pymysql://{USERNAME}:{PASSWORD}@{HOSTNAME}:{PORT}/{DATABASE}"
SQLALCHEMY_DATABASE_URI = DB_URI
```

（4）在项目的根目录下创建文件 models.py，代码如下：

```
# 资源包\Code\chapter3\3.13\1\flaskProject\models.py
from app import db
class User(db.Model):
    __tablename__ = "user"
    id = db.Column(db.Integer, primary_key = True, autoincrement = True)
    username = db.Column(db.String(50), nullable = False)
```

（5）在项目的根目录下创建文件 exts.py，代码如下：

```
# 资源包\Code\chapter3\3.13\1\flaskProject\exts.py
from flask_sqlalchemy import SQLAlchemy
db = SQLAlchemy()
```

（6）在项目的根目录下创建文件 manage.py，代码如下：

```
# 资源包\Code\chapter3\3.13\1\flaskProject\manage.py
from flask_script import Manager
from app import app
from exts import db
# MigrateCommand 中包含了 Migrate 的所有子命令
from flask_migrate import Migrate, MigrateCommand
from models import User
manager = Manager(app)
# 初始化 Migrate
Migrate(app, db)
manager.add_command('db', MigrateCommand)
if __name__ == '__main__':
    manager.run()
```

（7）打开命令提示符窗口，进入项目根目录下，激活虚拟环境，并输入命令 python manage.py db init，即可完成迁移仓库的初始化，如图 3-145 所示。

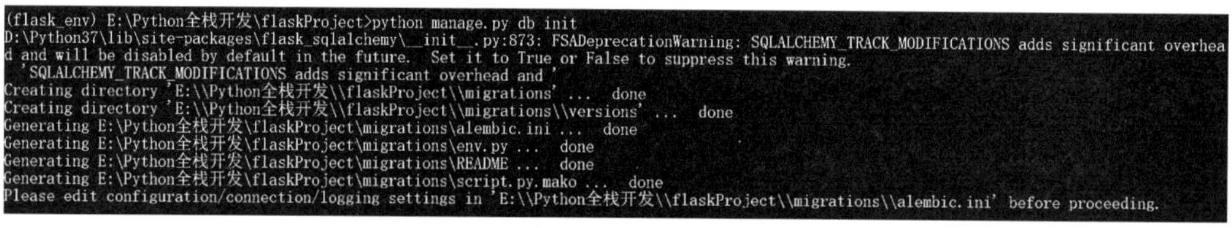

图 3-145 初始化迁移仓库

（8）输入命令 python manage.py db migrate，即可生成迁移脚本文件，如图 3-146 所示。

（9）输入命令 python manage.py db upgrade，即可完成数据库的更新，如图 3-147 所示。

（10）打开命令提示符窗口，登入 MySQL。

（11）在 MySQL 命令行窗口中输入 SQL 语句"desc user;"，其结果如图 3-148 所示。

图 3-146　生成迁移脚本文件

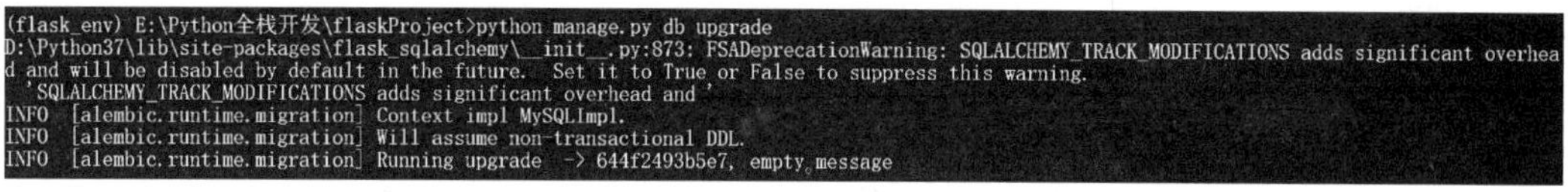

图 3-147　更新数据库

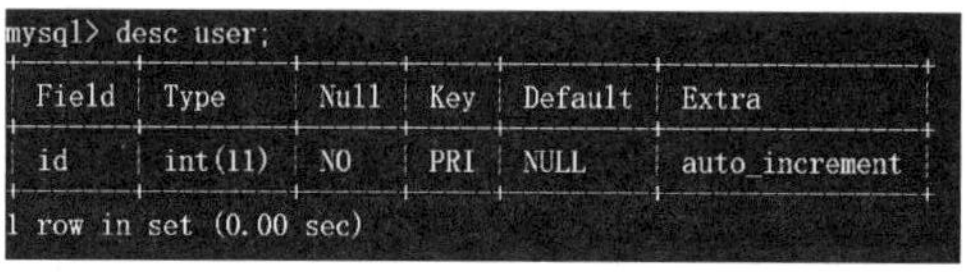

图 3-148　查看数据表结构

3.14　表单验证

在 Web 应用中，表单验证是一个基本而常见的任务，但是，由于 Flask 并没有提供全面的表单验证功能，所以在处理表单验证时会使代码复杂且混乱。

Python 的 WTForms 模块则通过提供表单的结构、验证和渲染等功能，简化了表单的处理流程。与此同时，Flask 的扩展 Flask-WTF 进一步整合了 WTForms，为开发者提供了更便捷、更灵活的表单处理方式。Flask-WTF 是建立在 WTForms 之上的 Flask 扩展，旨在简化 Web 应用中表单处理的流程，提供了与 Flask 框架的无缝集成，使表单的创建、验证和渲染变得非常容易。通过 Flask-WTF，开发者能够轻松地构建具有强大功能和良好用户体验的表单页面。

3.14.1　安装 WTForms 和 Flask-WTF

由于 WTForms 和 Flask-WTF 均属于 Python 的第三方库，所以需要进行安装，这里需要注意的是，WTForms 库不需要单独安装，只需在命令提示符中输入命令 pip install flask-wtf 便可同时完成两个库的安装。

3.14.2　HTML 表单验证

在 Flask 中，表单验证可以分为 6 步。

（1）导入 flask-wtf 模块、wtforms 模块和 wtforms.validators 模块中的相关类，其中，flask-wtf 模块包含 FlaskForm 类；wtforms 模块包含支持 HTML 的标准字段类（如表 3-13 所示）；wtforms.validators 模块包含对 HTML 表单进行验证的验证器类，包括内置验证器类和自定义验证器类。

表 3-13　WTForms 支持的 HTML 标准字段类

字　段　类	参　　数	描　　述
StringField	label,表示字段别名;	字符串字段
TextAreaField	validators,表示验证规则组成的列表;	多行文本字段
PasswordField		密码文本字段
HiddenField		隐藏文本字段
DateField	filters,表示过滤器列表,用于对提交的数据进行过滤;	日期字段
DateTimeField	description,表示描述的信息,通常用于生成帮助信息;	日期时间字段
IntegerField		整型字段
DecimalField	id,表示在字段的位置;	复数字段
FloatField	default,表示默认值;	浮点数字段
BooleanField	widget,表示 HTML 插件;	复选框字段
RadioField	render_kw,用于自定义 HTML 属性;	单选框字段
SelectField	choices,表示复选类型的选项	下拉列表字段
SelectMultipleField		多选下拉列表字段
FileField		文件上传字段
SubmitField		表单提交按钮

（2）创建表单类:表单类主要用于定义 HTML 表单中的字段,该类中的属性对应 HTML 表单中的每个字段。需要注意的是,表单类必须继承自 FlaskForm 类,并且其属性必须与表单字段中属性 name 的值一致。

（3）表单类实例化:在视图函数中,对表单类进行实例化操作。

（4）在模板中渲染表单:在 HTML 模板中,通过"表单类实例对象.字段名.label()"的方式对 HTML 表单字段的标签进行渲染,通过"表单类实例对象.字段名()"对控件进行渲染。

（5）处理验证数据:在视图函数中,通过表单类实例对象的 validate()方法处理数据验证之后的结果。

（6）验证表单数据:在表单类的字段定义中,可以通过内置验证器类（如表 3-14 所示）或自定义验证器类进行数据验证,其中,自定义验证器类需要在表单类中创建一种方法,并且该方法的命名规则需为"validate_字段名"。

表 3-14　内置验证器类

内置验证器类	参　　数	描　　述
Email	message,用于在验证失败时,自定义错误消息	验证是否是电子邮件地址
EqualTo	fieldname,该参数接受一个字符串,该字符串是需要比较的另一个字段的名称;message,用于在验证失败时,自定义错误消息	比较两个字段的值;常用于要求输入两次密钥进行确认的情况
Length	min,表示长度最小值;max,表示长度最大值;message,用于在验证失败时,自定义错误消息	验证输入字符串的长度
NumberRange	min,表示数值最小值;max,表示数值最大值;message,用于在验证失败时,自定义错误消息	验证输入的值在数字范围内
InputRequired	message,用于在验证失败时,自定义错误消息	验证该项数据为必填项,即要求非空

续表

内置验证器类	参　　数	描　　述
Regexp	regex,用于定义正则表达式；message,用于在验证失败时,自定义错误消息	使用正则表达式验证输入值
URL	message,用于在验证失败时,自定义错误消息	验证 URL
UUID	message,用于在验证失败时,自定义错误消息	验证数据是否是 UUID 类型

下面通过一个示例演示一下如何进行 HTML 表单验证。

（1）创建名为 flaskProject 的 Flask 项目。

（2）打开保存在项目根目录下的文件 app.py,代码如下：

```python
# 资源包\Code\chapter3\3.14\1\flaskProject\app.py
from flask import Flask, request, render_template
from forms import RegistForm
from flask_wtf import CSRFProtect
import os
app = Flask(__name__)
# CSRF 防御相关知识点,后续章节将为读者详细讲解
csrf = CSRFProtect()
csrf.init_app(app)
app.config['SECRET_KEY'] = os.urandom(24)
@app.route('/')
def hello_world():
    return '老夏学院!'
@app.route("/register/", methods = ["GET", "POST"])
def regist():
    if request.method == "GET":
        # (3)表单类实例化
        form = RegistForm(request.form)
        return render_template("register.html", form = form)
    else:
        form = RegistForm(request.form)
        # (5)处理验证数据
        if form.validate():
            return "注册成功!"
        else:
            print(form.errors)
            return "注册失败!"
if __name__ == '__main__':
    app.run(debug = True)
```

（3）在项目的根目录下的文件夹 templates 中创建模板文件 register.html,代码如下：

```html
# 资源包\Code\chapter3\3.14\1\flaskProject\templates\register.html
<!DOCTYPE html>
<html lang = "en">
    <head>
        <meta charset = "UTF-8">
        <title>注册页面</title>
        <style>
            .label {
                color: blue;
```

```html
            }
            .input {
                background-color: yellow;
            }
        </style>
    </head>
    <body>
        <form method="POST">
            <input type="hidden" name="csrf_token" value="{{ csrf_token() }}">
            <table>
                <tbody>
                    <tr>
                        {# (4)在模板中渲染表单 #}
                        {# <td>用户名:</td> #}
                        <td>{{ form.username.label(class = 'label') }}</td>
                        {# <td><input type="text" name="username"/></td> #}
                        <td>{{ form.username(class = 'input') }}</td>
                    </tr>
                    <tr>
                        <td>{{ form.password.label() }}</td>
                        <td><input type="password" name="password"/></td>
                    </tr>
                    <tr>
                        <td>确认密码:</td>
                        <td><input type="password" name="password_repeat"/></td>
                    </tr>
                    <tr>
                        <td>Email:</td>
                        <td><input type="text" name="email"/></td>
                    </tr>
                    <tr>
                        <td>年龄:</td>
                        <td><input type="text" name="age"/></td>
                    </tr>
                    <tr>
                        <td>电话:</td>
                        <td><input type="text" name="tel"/></td>
                    </tr>
                    <tr>
                        <td>个人主页:</td>
                        <td><input type="text" name="home_page"/></td>
                    </tr>
                    <tr>
                        <td>UUID:</td>
                        <td><input type="text" name="uuid"/></td>
                    </tr>
                    <tr>
                        <td>推荐人:</td>
                        <td><input type="text" name="reference"/></td>
                    </tr>
                    <tr>
                        <td>{{ form.tags.label }}</td>
                        <td>{{ form.tags() }}</td>
                    </tr>
                    <tr>
                        <td><input type="submit" value="立即注册"/></td>
```

```html
                <td>{{ form.remember.label }}{{ form.remember() }}</td>
            </tr>
        </tbody>
    </table>
  </form>
 </body>
</html>
```

（4）在项目的根目录下创建文件 forms.py，代码如下：

```python
#资源包\Code\chapter3\3.14\1\flaskProject\forms.py
#(1)导入 flask_wtf、wtforms 和 wtforms.validators 中所需的相关类
from flask_wtf import FlaskForm
from wtforms import StringField, IntegerField, ValidationError, BooleanField, SelectField
from wtforms.validators import Length, EqualTo, InputRequired, Email, NumberRange, Regexp, URL, UUID
#(2)创建表单类
class RegistForm(FlaskForm):
    #(3)验证表单数据
    username = StringField(label = '用户名',
                           validators = [Length(min = 3, max = 10, message = "用户名长度必须在3 到 10 位"), InputRequired()])
    password = StringField(label = '密码', validators = [Length(min = 6, max = 10, message = "密码长度必须在 6 到 10 位")])
    password_repeat = StringField(validators = [Length(min = 6, max = 10), EqualTo("password")])
    email = StringField(validators = [Email()])
    age = IntegerField(validators = [NumberRange(1, 130)])
    tel = StringField(validators = [Regexp(r'1[3456789]\d{9}')])
    home_page = StringField(validators = [URL()])
    uuid = StringField(validators = [UUID()])
    reference = StringField(validators = [Length(1, 20)])
    #自定义验证器
    def validate_reference(self, field):
        if field.data != "夏正东":
            raise ValidationError("推荐人错误")
    tags = SelectField("标签", choices = [("1", "Python"), ("2", "PHP"), ("3", "C++")])
    remember = BooleanField("记住用户名")
```

（5）运行上述程序，打开浏览器并在网址栏输入 http://127.0.0.1:5000/register/，其显示内容如图 3-149 所示。

（6）填入注册所需的相关信息，其显示内容如图 3-150 所示。

（7）由于填入的电话信息错误，导致无法通过表单验证，所以注册失败，其显示内容如图 3-151 所示。

3.14.3 文件上传验证

文件上传验证需要注意 3 点：一是表单类中对应属性的字段类型需为 FileField；二是其验证器需要使用 flask_wtf.file 模块中的 FileAllowed 类和 FileRequired 类，

图 3-149 浏览器中的显示内容

图 3-150　浏览器中的显示内容　　　　　　　图 3-151　浏览器中的显示内容

分别表示上传文件为必填项和上传允许的文件类型；三是在视图函数中，需要使用 werkzeug.datastructures 模块中的 CombinedMultiDict 类对 request.form 和 request.files 进行合并。

下面通过一个示例演示一下如何进行文件上传验证。

(1) 创建名为 flaskProject 的 Flask 项目。

(2) 打开保存在项目根目录下的文件 app.py，代码如下：

```
#资源包\Code\chapter3\3.14\2\flaskProject\app.py
from flask import Flask, request, render_template
from flask import send_from_directory
from werkzeug.datastructures import CombinedMultiDict
from forms import UploadFrom
import os
app = Flask(__name__)
UPLOAD_PATH = os.path.join(os.path.dirname(__file__), "upload")
@app.route('/')
def hello_world():
    return '老夏学院!'
@app.route("/register/", methods = ["GET", "POST"])
def regist():
    if request.method == "GET":
        return render_template("register.html")
    else:
        form = UploadFrom(CombinedMultiDict([request.form, request.files]))
        if form.validate():
            #pic = form.pic.data
            pic = request.files.get('pic')
            pic.save(os.path.join(UPLOAD_PATH, pic.filename))
            return '图片上传成功!'
        else:
            return '上传图片失败!'
#获取图片
```

```python
@app.route('/upload/<filename>/')
def get_pic(filename):
    # 该函数用于从指定目录将文件发送给客户端,其中,参数 directory 表示要发送文件的路径参
    # 数 filename,表示要发送的文件名
    return send_from_directory(directory = UPLOAD_PATH, filename = filename)
if __name__ == '__main__':
    app.run(debug = True)
```

(3) 在项目的根目录下创建文件 forms.py,代码如下:

```python
# 资源包\Code\chapter3\3.14\2\flaskProject\forms.py
from wtforms import Form, FileField
from flask_wtf.file import FileAllowed, FileRequired
class UploadFrom(Form):
    pic = FileField(validators = [FileRequired(), FileAllowed(['jpg', 'png', 'gif'])])
```

(4) 在项目的根目录下的文件夹 templates 中创建模板文件 register.html,代码如下:

```html
# 资源包\Code\chapter3\3.14\2\flaskProject\templates\register.html
<!DOCTYPE html>
<html lang = "en">
    <head>
        <meta charset = "UTF-8">
        <title>注册页面</title>
    </head>
    <body>
        <form action = "" method = "POST" enctype = "multipart/form-data">
            <table>
                <tbody>
                    <tr>
                        <td>上传图片:</td>
                        <td><input type = "file" name = "pic"/></td>
                    </tr>
                    <tr>
                        <td colspan = "2"><input type = "submit" value = "提交"/></td>
                    </tr>
                </tbody>
            </table>
        </form>
    </body>
</html>
```

(5) 在项目的根目录下创建文件夹 upload,用于接收上传的文件。

(6) 运行上述程序,打开浏览器并在网址栏输入 http://127.0.0.1:5000/register/,其显示内容如图 3-152 所示。

(7) 上传正确的图片文件,其显示内容如图 3-153 所示。

图 3-152　浏览器中的显示内容　　　　图 3-153　浏览器中的显示内容

（8）在浏览器中的网址栏输入 http://127.0.0.1:5000/upload/oldxia.png/，即可获取已上传的图片文件，其显示内容如图 3-154 所示。

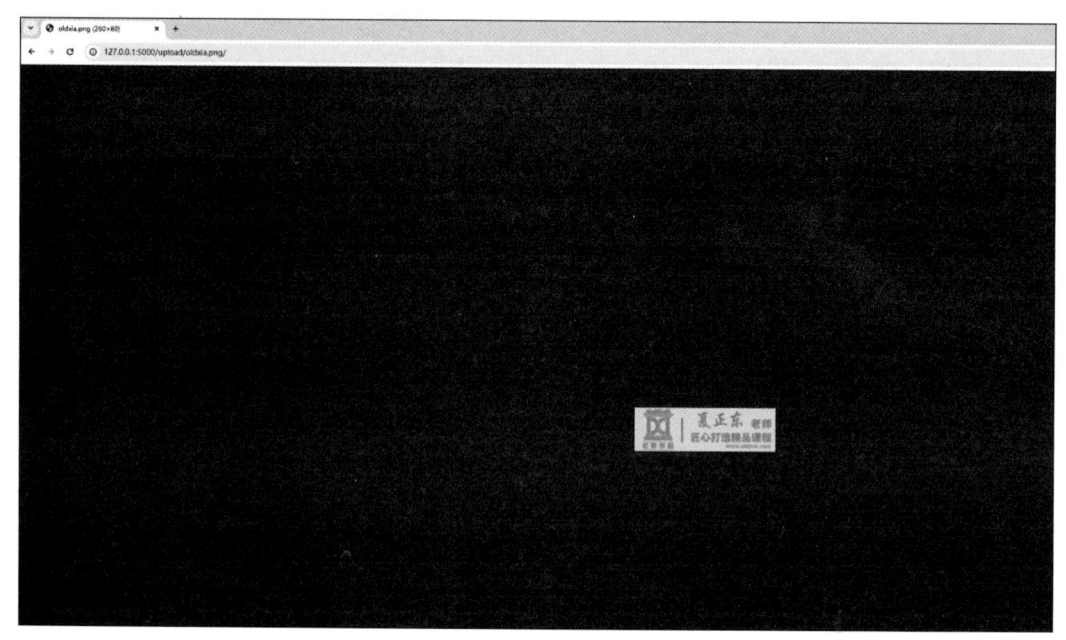

图 3-154　浏览器中的显示内容

3.15　Cookie 和 Session

关于 Cookie 和 Session 的具体概念，读者可以参考《Python 全栈开发——数据分析》一书中第 1 章的"模拟登录"部分内容，本书不再赘述。

3.15.1　设置、获取和删除 Cookie

1. 设置 Cookie

通过 Response 对象的 set_cookie()方法进行 Cookie 的设置，其语法格式如下：

```
set_cookie(key, value, max_age, expires, path, domain, secure, httponly)
```

其中，参数 key 表示 Cookie 的键，参数 value 表示 Cookie 的值，参数 max_age 表示 Cookie 被保存的时间，单位为秒，参数 expires 表示 Cookie 的具体过期时间，参数 path 用于限制 Cookie 的有效路径，参数 domain 用于设置 Cookie 可用的域名，参数 secure 用于设置 Cookie 是否仅通过 HTTPS 发送，参数 httponly 用于设置是否禁止 JavaScript 获取 Cookie。

2. 获取 Cookie

通过 request 对象的 cookies 属性来获取指定 Cookie 的值。

3. 删除 Cookie

通过 Response 对象的 delete_cookie()方法对 Cookie 进行删除，其语法格式如下：

```
delete_cookie(key)
```

其中,参数 key 表示 Cookie 的键。

下面通过一个示例演示一下如何设置、获取和删除 Cookie。

(1) 创建名为 flaskProject 的 Flask 项目。

(2) 打开保存在项目根目录下的文件 app.py,代码如下:

```python
# 资源包\Code\chapter3\3.15\1\flaskProject\app.py
from flask import Flask, Response, request
import datetime
app = Flask(__name__)
@app.route('/')
def index():
    return '老夏学院'
@app.route('/set_cookie/')
def set_cookie():
    res = Response('已设置 Cookie')

    # expires = datetime.datetime(year = 2024, month = 5, day = 1, hour = 12, minute = 0, second = 0)
    res.set_cookie('web_site', 'www.oldxia.com', max_age = 60)
    return res
@app.route('/get_cookie/')
def get_cookie():
    web_site = request.cookies.get('web_site')
    return web_site
@app.route('/del_cookie/')
def del_cookie():
    res = Response('已删除 Cookie')
    res.delete_cookie('web_site')
    return res
if __name__ == '__main__':
    app.run(debug = True)
```

(3) 运行上述程序,打开浏览器并在网址栏输入 http://127.0.0.1:5000/set_cookie/,即可设置 Cookie,其显示内容如图 3-155 所示。

图 3-155　浏览器中的显示内容

(4) 在浏览器中的网址栏输入 http://127.0.0.1:5000/get_cookie/,即可获取该 Cookie 的值,其显示内容如图 3-156 所示。

图 3-156　浏览器中的显示内容

(5) 在浏览器中的网址栏输入 http://127.0.0.1:5000/del_cookie/,即可删除该 Cookie,其显示内容如图 3-157 所示。

图 3-157　浏览器中的显示内容

3.15.2　设置、获取和删除 Session

为了防止恶意用户篡改 Session 中存储的数据,需要设置 Flask 配置项中的变量 SECRET_KEY,即为 Session 提供一个密钥,用于签名 Session 数据,以此来确保 Session 的完整性和安全性。

1. 设置 Session

通过 session 对象对 Session 进行设置。

此外,还可以通过 session 对象的 permanent 属性设置 Session 的过期时间,分两步:一是将 permanent 属性的值设置为 True;二是在 Flask 配置项中设置变量 PERMANENT_SESSION_LIFETIME 的值。需要注意的是,如果不设置变量 PERMANENT_SESSION_LIFETIME 的值,则默认过期时间为 1 个月。

2. 获取 Session

通过 session 对象的 get()方法来获取指定 Session 的值。

3. 删除 Session

删除 Session 有两种方式,一是通过 session 对象的 pop()方法删除指定的 Session;二是通过 session 对象的 clear()方法删除全部的 Session。

下面通过一个示例演示一下如何设置、获取和删除 Session。

(1) 创建名为 flaskProject 的 Flask 项目。

(2) 打开保存在项目根目录下的文件 app.py,代码如下:

```python
# 资源包\Code\chapter3\3.15\2\flaskProject\app.py
from flask import Flask, session
import os
import datetime
app = Flask(__name__)
# 设置 SECRET_KEY,并且应为一个复杂、随机的字符串
app.config['SECRET_KEY'] = os.urandom(24)
# 如果不设置具体的过期时间,则默认过期时间为 1 个月
app.config['PERMANENT_SESSION_LIFETIME'] = datetime.timedelta(hours=2)
@app.route('/')
def index():
    return '老夏学院'
@app.route('/set_session/')
def set_session():
    session['web_site'] = 'www.oldxia.com'
    session.permanent = True
    return '已设置 Session'
@app.route('/get_session/')
def get_session():
    web_site = session.get('web_site')
    return web_site
@app.route('/del_session/')
def del_session():
```

```
        session.pop('web_site')
        return '已删除 Session'
if __name__ == '__main__':
    app.run(debug = True)
```

（3）运行上述程序，打开浏览器并在网址栏输入 http://127.0.0.1:5000/set_session/，即可设置 Session，其显示内容如图 3-158 所示。

图 3-158　浏览器中的显示内容

（4）在浏览器中的网址栏输入 http://127.0.0.1:5000/get_session/，即可获取该 Session 的值，其显示内容如图 3-159 所示。

图 3-159　浏览器中的显示内容

（5）在浏览器中的网址栏输入 http://127.0.0.1:5000/del_session/，即可删除该 Session，其显示内容如图 3-160 所示。

图 3-160　浏览器中的显示内容

3.16　CSRF 防御

CSRF（Cross-Site Request Forgery，跨站请求伪造）是一种挟制用户在当前已登录的 Web 应用程序上执行非本意的操作的攻击方法。攻击者通过 HTTP 请求将数据发送到服务器，进而盗取 Cookie。在盗取到 Cookie 之后，攻击者不仅可以获取用户的相关信息，还可以修改该 Cookie 所关联的账户信息。

在 Flask 中，flask_wtf 模块提供了一套基于 Token 校验的完善的 CSRF 防护体系，可以非常简单地解决 CSRF 攻击问题，具体可以分为 4 步。

（1）导入 CSRFProtect 类，并实例化。

（2）通过 CSRFProtect 实例对象的 init_app() 方法进行初始化，与 Flask 实例对象进行绑定。

（3）设置 Flask 配置项中变量 SECRET_KEY 的值，用于加密生成 CSRF 令牌中的值。

（4）在对应的视图文件中添加 CSRF 令牌。

下面通过一个示例演示一下如何进行 CSRF 防御。

（1）创建名为 flaskProject 的 Flask 项目。

（2）打开保存在项目根目录下的文件 app.py，代码如下：

```python
# 资源包\Code\chapter3\3.16\1\flaskProject\app.py
from flask import Flask, render_template, session
from flask_wtf import CSRFProtect
import os
app = Flask(__name__)
csrf = CSRFProtect()
csrf.init_app(app)
app.config['SECRET_KEY'] = os.urandom(24)
@app.route("/register/")
def user():
    return render_template("register.html")
@app.route("/transfer/", methods=["POST"])
def transfer():
    csrf_token = session.get("csrf_token")
    return f"转账成功!csrf_token 的值为{csrf_token}"
if __name__ == '__main__':
    app.run(debug=True)
```

（3）在项目的根目录下的文件夹 templates 中创建模板文件 register.html，代码如下：

```html
# 资源包\Code\chapter3\3.16\1\flaskProject\templates\register.html
<!doctype html>
<html lang="en">
    <head>
        <meta charset="UTF-8">
        <title>Document</title>
    </head>
    <body>
        <form action="{{ url_for('transfer') }}" method="post">
            <input type="hidden" name="csrf_token" value="{{ csrf_token() }}">
            账号:<input type="text" name="username"><br><br>
            密码:<input type="password" name="password"><br><br>
            <input type="submit" value="转账">
        </form>
    </body>
</html>
```

（4）运行上述程序，打开浏览器并在网址栏输入 http://127.0.0.1:5000/register/，其显示内容如图 3-161 所示。

（5）输入任意的账号和密码，然后单击"转账"按钮，模拟登录账户操作，其显示内容如图 3-162 所示。

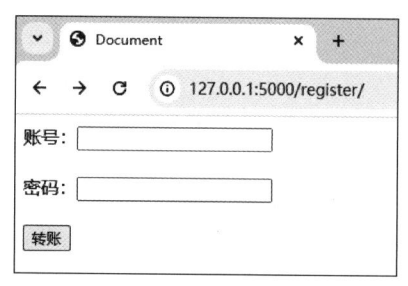

图 3-161　浏览器中的显示内容　　　　　　图 3-162　浏览器中的显示内容

3.17 上下文

Flask 上下文指的是在代码执行到某一行时，根据之前代码已执行的操作，以及下文即将执行的逻辑，来决定当前可以使用的变量或者可以完成的事情。

Flask 上下文相当于一个容器，其保存了程序在运行的过程中的一些信息，例如变量、函数、类与对象等。

在 Flask 中，包括两种上下文，即应用上下文和请求上下文。

3.17.1 应用上下文

当一个 Flask 应用启动时会自动创建一个应用上下文，其表示整个应用的运行环境，用于存储应用全局的变量和配置，包括应用配置、日志器和数据库连接信息等。

这里需要注意的是，对于每个请求来讲，应用上下文会在当前请求处理之前创建，并且会一直存在到请求处理完毕后才被销毁。这就意味着，应用上下文可以在整个请求生命周期内共享数据。

应用上下文提供的对象包括 current_app 对象和 g 对象。

下面通过一个示例演示一下如何使用应用上下文。

（1）创建名为 flaskProject 的 Flask 项目。

（2）打开保存在项目根目录下的文件 app.py，代码如下：

```python
#资源包\Code\chapter3\3.17\1\flaskProject\app.py
from flask import Flask, current_app
app = Flask(__name__)
with app.app_context():
    print(current_app.name)
@app.route('/')
def hello_world():
    return '老夏学院!'
if __name__ == '__main__':
    app.run(debug = True)
```

（3）运行上述程序，PyCharm 中的输出结果如图 3-163 所示。

```
* Environment: production
  WARNING: This is a development server. Do not use it in a production deployment.
  Use a production WSGI server instead.
* Debug mode: off
app
```

图 3-163　PyCharm 中的输出结果

在 Flask 中，应用上下文所提供的 g 对象是一个全局对象，可以在整个应用程序中使用，并在每个请求期间自动创建和销毁。

g 对象本质上是一个轻量级的容器，可以用来存储应用程序中的任意数据，这些数据既可以是请求特定的，也可以是跨请求共享的。

g 对象的使用方式也非常简单，通过 g 对象的属性即可进行数据存储。

下面通过一个示例演示一下如何使用 g 对象。

(1) 创建名为 flaskProject 的 Flask 项目。
(2) 打开保存在项目根目录下的文件 app.py，代码如下：

```python
# 资源包\Code\chapter3\3.17\2\flaskProject\app.py
from flask import Flask, request, g
app = Flask(__name__)
def is_admin():
    if g.username == 'oldxia':
        return True
    else:
        return False
@app.route('/login/')
def login():
    username = request.args.get('username')
    g.username = username
    if is_admin():
        return 'Login is OK'
    else:
        return 'Sorry'
if __name__ == '__main__':
    app.run(debug = True)
```

(3) 运行上述程序，打开浏览器并在网址栏输入 http://127.0.0.1:5000/login/?username=oldxia，其显示内容如图 3-164 所示。

图 3-164　浏览器中的显示内容

(4) 在浏览器中的网址栏输入 http://127.0.0.1:5000/login/?username=admin，其显示内容如图 3-165 所示。

图 3-165　浏览器中的显示内容

3.17.2　请求上下文

当请求到达 Flask 应用时，每个请求都会有一个专属的请求上下文，用于存储请求相关的变量和信息，包括请求路径、请求方法、请求参数和会话信息等。

请求上下文提供的对象包括 request 对象和 session 对象。

这里需要注意的是，在 Flask 中，可以直接在视图函数中使用 request 对象获取相关的数据。

下面通过一个示例演示一下如何使用请求上下文。

(1) 创建名为 flaskProject 的 Flask 项目。
(2) 打开保存在项目根目录下的文件 app.py,代码如下:

```python
# 资源包\Code\chapter3\3.17\3\flaskProject\app.py
from flask import Flask, current_app, url_for
app = Flask(__name__)
with app.app_context():
    print(current_app.name)
@app.route('/')
def index():
    return '老夏学院!'
@app.route('/web_site/')
def web_site():
    return 'http://www.oldxia.com'
with app.test_request_context():
    print(url_for('web_site'))
if __name__ == '__main__':
    app.run(debug = True)
```

(3) 运行上述程序,PyCharm 中的输出结果如图 3-166 所示。

```
* Environment: production
  WARNING: This is a development server. Do not use it in a production deployment.
  Use a production WSGI server instead.
* Debug mode: off
app
/web_site/
```

图 3-166　PyCharm 中的输出结果

3.17.3　应用上下文和请求上下文的区别

首先,应用上下文是全局的,表示整个 Flask 应用的运行环境,而请求上下文则是针对每个请求独立存在的,表示该请求的运行环境。

其次,在应用程序的整个生命周期中,应用上下文只有一个,而且存在于应用的整个生命周期之中,而请求上下文则会随着请求动态地创建或销毁。

最后,对于每个请求来讲,应用上下文是每个请求共享的,而请求上下文则是每个请求所独有的。

下面通过一个例子来说明应用上下文和请求上下文的区别。

有一家大型超市(Flask 应用),每当有顾客到超市购买东西时(请求到达 Flask 应用),店员就会为每名顾客分配一个购物车(本地线程)。

这个购物车中存放了每名顾客的商品、折扣券(请求上下文)等,并且购物车里面的东西是每名顾客独有的(不同请求的请求上下文是独立的)。

当顾客消费完毕后,购物车里面的东西就会被清空,店员就会回收这些购物车,等待分配给下一名顾客使用(请求处理完毕后,就会将请求上下文从当前线程中删除并销毁,并为后续新到的请求分配线程)。

而这个超市里面的电梯、货架及商品(应用上下文)都是每名顾客所共享的(应用上下文是每个请求共享的)。

当超市关门时,电梯、货架及商品就不能对外使用了(应用上下文存在于应用的生命周期中)。

3.18 钩子函数

在 Flask 中,钩子函数指的是在执行函数和目标函数之间挂载的函数,其通过特定的函数装饰器实现。

1. @Flask 实例对象.before_first_request

该钩子函数用于在处理第 1 个请求之前执行。

下面通过一个示例演示一下如何使用钩子函数 before_first_request。

(1) 创建名为 flaskProject 的 Flask 项目。

(2) 打开保存在项目根目录下的文件 app.py,代码如下:

```python
# 资源包\Code\chapter3\3.18\1\flaskProject\app.py
from flask import Flask
app = Flask(__name__)
@app.route('/')
def index():
    print('执行 index 函数!')
    return '老夏学院!'
@app.before_first_request
def before_first_request():
    print("开始启动程序")
if __name__ == '__main__':
    app.run(debug = True)
```

(3) 运行上述程序,打开浏览器并在网址栏输入 http://127.0.0.1:5000/。此时,PyCharm 中的输出结果如图 3-167 所示。

(4) 再次在浏览器中的网址栏输入 http://127.0.0.1:5000/。此时,PyCharm 中的输出结果如图 3-168 所示。

图 3-167 PyCharm 中的输出结果　　图 3-168 PyCharm 中的输出结果

2. @Flask 实例对象.before_request

该钩子函数用于在处理每次请求之前执行。

下面通过一个示例演示一下如何使用钩子函数 before_request。

(1) 创建名为 flaskProject 的 Flask 项目。

(2) 打开保存在项目根目录下的文件 app.py,代码如下:

```python
# 资源包\Code\chapter3\3.18\2\flaskProject\app.py
from flask import Flask
app = Flask(__name__)
@app.route('/')
def index():
    print('执行 index 函数!')
```

```
        return '老夏学院!'
@app.before_request
def before_request():
    print("开始执行API")
if __name__ == '__main__':
    app.run(debug = True)
```

(3) 运行上述程序,打开浏览器并在网址栏输入 http://127.0.0.1:5000/。此时,PyCharm 中的输出结果如图 3-169 所示。

(4) 再次在浏览器中的网址栏输入 http://127.0.0.1:5000/。此时,PyCharm 中的输出结果如图 3-170 所示。

图 3-169　PyCharm 中的输出结果　　图 3-170　PyCharm 中的输出结果

3．@Flask 实例对象.after_request

该钩子函数用于在处理每次请求之后执行,需要注意的是,其需要传递一个 Response 对象作为参数。

下面通过一个示例演示一下如何使用钩子函数 after_request。

(1) 创建名为 flaskProject 的 Flask 项目。

(2) 打开保存在项目根目录下的文件 app.py,代码如下:

```
# 资源包\Code\chapter3\3.18\3\flaskProject\app.py
from flask import Flask
app = Flask(__name__)
@app.after_request
def after_request(response):
    response.headers['web_site'] = 'http://www.oldxia.com'
    return response
@app.route('/')
def index():
    return '老夏学院'
if __name__ == '__main__':
    app.run()
```

(3) 运行上述程序,打开浏览器并右击,选择"检查"。

(4) 在浏览器中的网址栏输入 http://127.0.0.1:5000/,其显示内容如图 3-171 所示。

(5) 此时,查看 Network 选项,可以看到设置的响应头,其显示内容如图 3-172 所示。

图 3-171　浏览器中的显示内容

图 3-172　浏览器中的显示内容

4. @Flask 实例对象.context_processor

该钩子函数为上下文处理器，并且必须返回一个字典，其主要功能是，当变量在很多模板中需要被使用时，可以通过该钩子函数返回，而无须使用每个视图函数中的 render_template()方法进行传递，这样可以使代码更加简洁，便于维护。

下面通过一个示例演示一下如何使用钩子函数 context_processor。

（1）创建名为 flaskProject 的 Flask 项目。

（2）打开保存在项目根目录下的文件 app.py，代码如下：

```python
# 资源包\Code\chapter3\3.18\4\flaskProject\app.py
from flask import Flask, g, session, render_template
import os
app = Flask(__name__)
app.config['SECRET_KEY'] = os.urandom(24)
@app.route('/')
def index():
    print('执行index函数！')
    session['username'] = 'oldxia'
    return render_template('index.html')
@app.route('/home/')
def home():
    print('进入个人中心')
    if hasattr(g, 'username'):
        print('username:', g.username)
    return render_template('index.html')
@app.before_request
def before_request():
    print('before_request!')
    username = session.get('username')
    if username:
        g.username = username
@app.context_processor
def context_processor():
    if hasattr(g, 'username'):
        return {'current_user': g.username}
    else:
        return {}
if __name__ == '__main__':
    app.run(debug=True)
```

（3）在项目的根目录下的文件夹 templates 中创建模板文件 index.html，代码如下：

```html
# 资源包\Code\chapter3\3.18\4\flaskProject\templates\index.html
<!DOCTYPE html>
<html lang="en">
    <head>
        <meta charset="UTF-8">
        <meta http-equiv="X-UA-Compatible" content="IE=edge">
        <meta name="viewport" content="width=device-width, initial-scale=1.0">
        <title>Document</title>
    </head>
    <body>
        <h1>这个是个人主页</h1>
        <p>当前用户：{{ current_user }}</p>
```

```
        </body>
</html>
```

(4) 运行上述程序,打开浏览器并在网址栏输入 http://127.0.0.1:5000/home/,其显示内容如图 3-173 所示。

(5) 在浏览器中的网址栏输入 http://127.0.0.1:5000/,其显示内容如图 3-174 所示。

图 3-173　浏览器中的显示内容(1)

图 3-174　浏览器中的显示内容(2)

(6) 在浏览器中的网址栏输入 http://127.0.0.1:5000/home/,其显示内容如图 3-175 所示。

5. @Flask 实例对象.errorhandler

该钩子函数用于接收状态码,并可以自定义返回当前状态码的响应处理方法。这里需要注意的是,该钩子函数必须传递一个参数,用于接收错误信息,否则会报错。

图 3-175　浏览器中的显示内容(3)

此外,还可以通过 abort() 函数手动抛出相应的错误。

下面通过一个示例演示一下如何使用钩子函数 errorhandler。

(1) 创建名为 flaskProject 的 Flask 项目。

(2) 打开保存在项目根目录下的文件 app.py,代码如下:

```
# 资源包\Code\chapter3\3.18\5\flaskProject\app.py
from flask import Flask, render_template, abort
app = Flask(__name__)
@app.route('/')
def index():
    print('执行 index 函数!')
    abort(500)
    return render_template('index.html')
@app.errorhandler(500)
def server_error(error):
    # 此处推荐写上状态码,以便明确告知服务器错误类型
    return render_template('500.html'), 500
if __name__ == '__main__':
    app.run(debug = True)
```

(3) 在项目的根目录下的文件夹 templates 中创建模板文件 index.html,代码如下:

```
# 资源包\Code\chapter3\3.18\5\flaskProject\templates\index.html
<!DOCTYPE html>
<html lang = "en">
    <head>
        <meta charset = "UTF-8">
        <title>Document</title>
    </head>
    <body>
        <h1>这是首页!</h1>
    </body>
</html>
```

（4）在项目的根目录下的文件夹 templates 中创建模板文件 500.html，代码如下：

```
# 资源包\Code\chapter3\3.18\5\flaskProject\templates\500.html
<!DOCTYPE html>
<html lang = "en">
    <head>
        <meta charset = "UTF-8">
        <title>500 错误</title>
    </head>
    <body>
        <h1>内部服务器错误,请勿频繁刷新</h1>
    </body>
</html>
```

（5）运行上述程序，打开浏览器并在网址栏输入 http://127.0.0.1:5000/，其显示内容如图 3-176 所示。

图 3-176　浏览器中的显示内容

3.19　信号

Flask 信号是一种事件机制，其可以让应用程序在特定事件发生时自动地将一个通知发送给所有注册了该事件的处理函数。

此外，开发者可以在应用程序中注册一个或多个处理函数，当信号被触发时，这些处理函数就会被自动调用，用于执行相关操作，例如记录日志、发送邮件或更新数据库等。

使用 Flask 信号的优点是可以将应用程序的不同部分解耦，使其更加模块化和灵活，即通过注册不同的信号处理函数，可以将应用程序的不同功能分离开，方便维护和扩展。此外，信号还可以帮助开发者实现更高级的功能，例如异步任务管理、缓存更新等。

这里需要注意的是，信号本身不具有返回值，因为它会影响原有流程的执行。

3.19.1　信号的安装

在 Flask 中,由于信号基于 blinker 模块,并且该模块属于 Python 的第三方库,所以需要进行安装,只需在命令提示符中输入命令 pip install blinker。

3.19.2　自定义信号

自定义信号分为以下 3 步。

(1) 创建信号:首先通过 blinker 模块中的 Namespace 类创建命名空间,然后使用该命名空间的 signal 类创建信号对象。

(2) 监听信号:通过信号对象的 connect()方法监听信号,并且该方法需要传递一个参数,用于注册对应的处理函数。这里需要注意的是,该处理函数必须具有一个参数 sender,用于接收发送者的信号。

(3) 发送信号:通过信号对象的 send()方法发送信号。

下面通过一个示例演示一下如何自定义信号。

(1) 创建名为 flaskProject 的 Flask 项目。

(2) 打开保存在项目根目录下的文件 app.py,代码如下:

```python
# 资源包\Code\chapter3\3.19\1\flaskProject\app.py
from flask import Flask, request
from signals import login_signal
app = Flask(__name__)
@app.route('/')
def hello_world():
    return '老夏学院!'
@app.route('/login/')
def login():
    username = request.args.get('username')
    if username:
        # 发送信号
        login_signal.send(username=username)
        return f'您好{username},欢迎登录'
    else:
        return '登录失败,请输入用户名!'
if __name__ == '__main__':
    app.run(debug=True)
```

(3) 在项目的根目录下创建文件 signals.py,代码如下:

```python
# 资源包\Code\chapter3\3.19\1\flaskProject\signals.py
from blinker import Namespace
from datetime import datetime
from flask import request
# 创建信号
namespace = Namespace()
login_signal = namespace.signal('login')
# 监听信号
def login_log(sender, username):
    username = username
```

```python
    time = datetime.now()
    ip = request.remote_addr
    log = f"登录日志:用户名:{username},登录时间:{time},IP地址:{ip}\n"
    with open('login_log.txt', 'a+', encoding='utf-8') as f:
        f.write(log)
login_signal.connect(login_log)
```

(4) 运行上述程序,打开浏览器并在网址栏输入 http://127.0.0.1:5000/login/,其显示内容如图 3-177 所示。

(5) 在浏览器中的网址栏输入 http://127.0.0.1:5000/login/?username=oldxia,其显示内容如图 3-178 所示。

图 3-177　浏览器中的显示内容

图 3-178　浏览器中的显示内容

(6) 打开在项目根目录下所创建的文件 login_log.txt,其内容如图 3-179 所示。

图 3-179　文件 login_log.txt 中的内容

3.19.3　内置信号

在 Flask 中,内置了多种信号,用于处理对应的事件。

1. before_render_template

该信号用于在模板渲染之前执行,通过该信号可以修改要渲染的模板或添加数据。

下面通过一个示例,演示一下如何使用内置信号 before_render_template。

(1) 创建名为 flaskProject 的 Flask 项目。

(2) 打开保存在项目根目录下的文件 app.py,代码如下:

```python
# 资源包\Code\chapter3\3.19\2\flaskProject\app.py
from flask import Flask, before_render_template, render_template
app = Flask(__name__)
@app.route('/')
def index():
    print('开始渲染模板')
    return render_template('index.html')
def before_template(sender, template, context):
    print(sender)
    print(template)
    print(context)
before_render_template.connect(before_template)
if __name__ == '__main__':
    app.run(debug=True)
```

(3) 在项目的根目录下的文件夹 templates 中创建模板文件 index.html，代码如下：

```
#资源包\Code\chapter3\3.19\2\flaskProject\templates\index.html
<!DOCTYPE html>
<html lang="en">
    <head>
        <meta charset="UTF-8">
        <title>Title</title>
    </head>
    <body>
        老夏学院
    </body>
</html>
```

(4) 运行上述程序，打开浏览器并在网址栏输入 http://127.0.0.1:5000/。此时，PyCharm 中的输出结果如图 3-180 所示。

```
开始渲染模板
<Flask 'app'>
<Template 'index.html'>
{'g': <flask.g of 'app'>, 'request': <Request 'http://127.0.0.1:5000/' [GET]>, 'session': <NullSession {}>}
127.0.0.1 - - [16/Apr/2024 20:43:56] "GET / HTTP/1.1" 200 -
```

图 3-180　PyCharm 中的输出结果

2. template_rendered

该信号用于在模板渲染之后执行，通过该信号可以进行一些后处理操作。

下面通过一个示例演示一下如何使用内置信号 template_rendered。

(1) 创建名为 flaskProject 的 Flask 项目。

(2) 打开保存在项目根目录下的文件 app.py，代码如下：

```
#资源包\Code\chapter3\3.19\3\flaskProject\app.py
from flask import Flask, template_rendered, render_template
app = Flask(__name__)
@app.route('/')
def index():
    print('开始渲染模板')
    return render_template('index.html')
def after_template(sender, template, context):
    print(sender)
    print(template)
    print(context)
template_rendered.connect(after_template)
if __name__ == '__main__':
    app.run(debug=True)
```

(3) 在项目的根目录下的文件夹 templates 中创建模板文件 index.html，代码如下：

```
#资源包\Code\chapter3\3.19\3\flaskProject\templates\index.html
<!DOCTYPE html>
<html lang="en">
    <head>
        <meta charset="UTF-8">
        <title>Title</title>
```

```
        </head>
        <body>
            老夏学院
        </body>
</html>
```

（4）运行上述程序，打开浏览器并在网址栏输入 http://127.0.0.1:5000/。此时，PyCharm 中的输出结果如图 3-181 所示。

```
开始渲染模板
<Flask 'app'>
<Template 'index.html'>
{'g': <flask.g of 'app'>, 'request': <Request 'http://127.0.0.1:5000/' [GET]>, 'session': <NullSession {}>}
```

图 3-181　PyCharm 中的输出结果

3．got_request_exception

该信号用于在请求执行出现异常时执行，通过该信号可以处理请求异常并记录错误信息。

下面通过一个示例演示一下如何使用内置信号 got_request_exception。

（1）创建名为 flaskProject 的 Flask 项目。

（2）打开保存在项目根目录下的文件 app.py，代码如下：

```python
# 资源包\Code\chapter3\3.19\4\flaskProject\app.py
from flask import Flask, got_request_exception
app = Flask(__name__)
@app.route('/')
def index():
    val = 1 / 0
    return '老夏学院'
def request_exception_log(sender, exception):
    print(sender)
    print(exception)
got_request_exception.connect(request_exception_log)
if __name__ == '__main__':
    app.run(debug=True)
```

（3）运行上述程序，打开浏览器并在网址栏输入 http://127.0.0.1:5000/。此时，PyCharm 中的输出结果如图 3-182 所示。

```
<Flask 'app'>
division by zero
[2024-04-16 15:02:30,704] ERROR in app: Exception on / [GET]
```

图 3-182　PyCharm 中的输出结果

4．request_started

该信号用于在请求到来之前执行，通过该信号可以执行初始化操作或日志记录。

下面通过一个示例，演示一下如何使用内置信号 request_started。

（1）创建名为 flaskProject 的 Flask 项目。

（2）打开保存在项目根目录下的文件 app.py，代码如下：

```
#资源包\Code\chapter3\3.19\5\flaskProject\app.py
from flask import Flask, request_started
app = Flask(__name__)
@app.route('/')
def index():
    print('index_page')
    return '老夏学院'
def before_request(sender):
    print(sender)
request_started.connect(before_request)
if __name__ == '__main__':
    app.run(debug = True)
```

(3) 运行上述程序,打开浏览器并在网址栏输入 http://127.0.0.1:5000/。此时,PyCharm 中的输出结果如图 3-183 所示。

```
<Flask 'app'>
index_page
```

图 3-183　PyCharm 中的输出结果

5. request_finished

该信号用于在请求结束后执行,通过该信号可以进行清理操作或处理请求完成后的逻辑。

下面通过一个示例演示一下如何使用内置信号 request_finished。

(1) 创建名为 flaskProject 的 Flask 项目。

(2) 打开保存在项目根目录下的文件 app.py,代码如下:

```
#资源包\Code\chapter3\3.19\6\flaskProject\app.py
from flask import Flask, request_finished
app = Flask(__name__)
@app.route('/')
def index():
    print('index_page')
    return '老夏学院'
def after_request(sender, response):
    print(sender)
    print(response)
request_finished.connect(after_request)
if __name__ == '__main__':
    app.run(debug = True)
```

(3) 运行上述程序,打开浏览器并在网址栏输入 http://127.0.0.1:5000/。此时,PyCharm 中的输出结果如图 3-184 所示。

```
index_page
<Flask 'app'>
<Response 12 bytes [200 OK]>
```

图 3-184　PyCharm 中的输出结果

第 4 章 Flask 项目实战：网上图书商城

本章将学习如何实现网上图书商城，以便于更好地理解 Flask 的相关使用方式。

4.1 程序概述

1. 登录页面

登录页面主要用于客户账号和客户密码的输入，其界面如图 4-1 所示。

图 4-1 登录页面

2. 注册页面

注册页面主要用于客户注册登录所需的账号，其界面如图 4-2 所示。

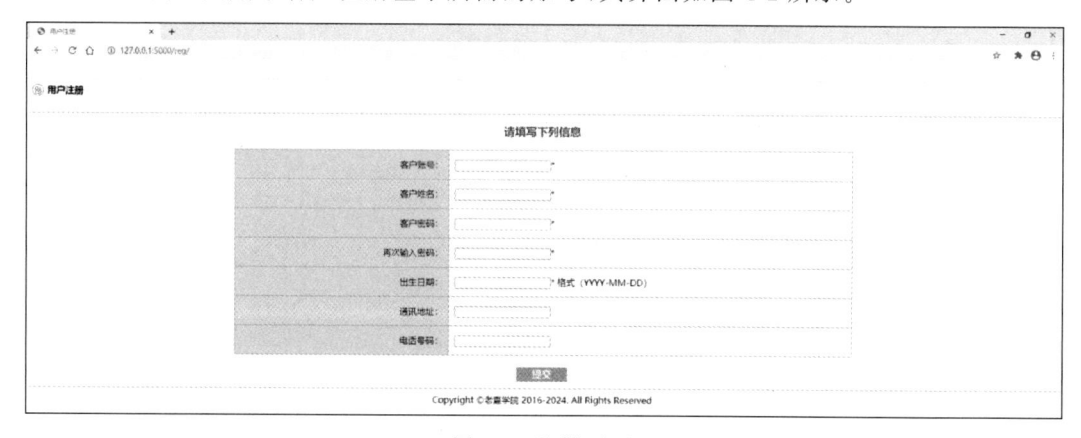

图 4-2 注册页面

3. 主页面

主页面主要用于展示商品列表页面和购物车页面的导航，其界面如图 4-3 所示。

图 4-3　主页面

4. 商品列表页面

商品列表页面主要用于显示全部商品，其界面如图 4-4 所示。

图 4-4　商品列表页面

5. 商品详情页面

商品详情页面主要用于显示商品的详细信息，其界面如图 4-5 所示。

图 4-5　商品详情页面

6. 购物车页面

购物车页面主要用于显示所购商品的订单信息，其界面如图 4-6 所示。

图 4-6　购物车页面

7. 账户信息页面

账户信息页面主要用于显示当前客户的账户信息,其界面如图 4-7 所示。

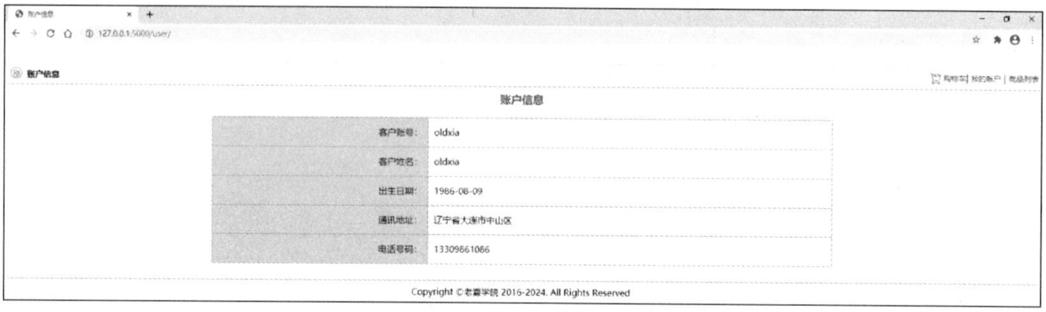

图 4-7　账户信息页面

4.2　创建数据库

该项目所使用的数据库为 SQLite,创建数据库的 SQL 语句如下:

```sql
#资源包\BookStore\db\store-schema.sql
drop table if exists Customers;
drop table if exists OrderLineItems;
drop table if exists Goods;
drop table if exists Orders;
/* ============================================================ */
/* Table: Customers */
/* ============================================================ */
create table Customers
(
   id         varchar(20) primary key,
   name       varchar(50) not null,
   password varchar(20) not null,
   address    varchar(100),
   phone      varchar(20),
   birthday varchar(20)
);
/* ============================================================ */
/* Table: Goods */
/* ============================================================ */
create table Goods
(
   id                integer primary key autoincrement,
   name              varchar(100) not null,
   author            varchar(30),
   press             varchar(200),
   isbn              varchar(30),
   edition           varchar(30),
   packaging         varchar(30),
   format            varchar(30),
   publication_time varchar(30),
   paper             varchar(30),
   price             varchar(30),
```

```
    description         varchar(200),
    image               varchar(100)
);
/* ============================================================ */
/* Table: Orders */
/* ============================================================ */
create table Orders
(
    id              varchar(20) primary key,
    order_date      varchar(20),
    status          integer default 1,
    total           float
);
/* ============================================================ */
/* Table: OrderLineItems */
/* ============================================================ */
create table OrderLineItems
(
    id          integer primary key autoincrement,
    goodsid     integer not null references Goods (id),
    orderid     integer not null references Orders (id),
    quantity    integer,
    sub_total   float
);
```

上面的 SQL 语句创建了 4 个数据表,即客户表、商品表、订单表和详细订单表。

1. 客户表

该表主要用于存储用户的相关信息,包括客户账号、客户姓名、客户密码、通信地址、电话号码和出生日期,其表结构如表 4-1 所示。

表 4-1 客户表的表结构

字 段 名	数据类型	长 度	主 键	外 键	备 注
id	varchar(20)	20	YES	NO	客户账号
name	varchar(50)	50	NO	NO	客户姓名
password	varchar(20)	20	NO	NO	客户密码
address	varchar(100)	100	NO	NO	通信地址
phone	varchar(20)	20	NO	NO	电话号码
birthday	varchar(20)	20	NO	NO	出生日期

2. 商品表

该表主要用于存储商品的相关信息,包括图书名称、作者、出版社、ISBN、版次、包装、开本、出版时间、用纸、价格、图书详细描述和图书图片,其表结构如表 4-2 所示。

表 4-2 商品表的表结构

字 段 名	数据类型	长 度	主 键	外 键	备 注
id	integer	—	YES	NO	图书 id
name	varchar(100)	100	NO	NO	图书名称
author	varchar(30)	30	NO	NO	作者
press	varchar(200)	200	NO	NO	出版社

续表

字 段 名	数据类型	长 度	主 键	外 键	备 注
isbn	varchar(30)	30	NO	NO	ISBN
edition	varchar(30)	30	NO	NO	版次
packaging	varchar(30)	30	NO	NO	包装
format	varchar(30)	30	NO	NO	开本
publication_time	varchar(30)	30	NO	NO	出版时间
paper	varchar(30)	30	NO	NO	用纸
price	varchar(30)	30	NO	NO	价格
description	varchar(200)	200	NO	NO	图书详细描述
image	varchar(100)	100	NO	NO	图书图片

此外，向该表中插入数据的 SQL 语句如下：

```
#资源包\BookStore\db\store-dataload.sql
insert into Goods values (1,'Python全栈开发——基础入门','夏正东','清华大学出版社',
'9787302600909','1','平装','16开','2022-07-01','胶版纸','79','Python全栈开发——基础入门
(清华开发者书库.Python)作者：夏正东 Python畅销书籍','dc670c75c629548c.jpg');
insert into Goods values (2,'Python全栈开发——高阶编程','夏正东','清华大学出版社',
'9787302608943','1','平装','16开','2022-08-01','胶版纸','89','Python全栈开发——高阶编程
(清华开发者书库.Python)作者：夏正东 Python畅销书籍','14ce4d2acd51eed8.jpg');
insert into Goods values (3,'Python全栈开发——数据分析','夏正东','清华大学出版社',
'9787302625001','1','平装','16开','2023-03-01','胶版纸','79','Python全栈开发——数据分析
(清华开发者书库.Python)作者：夏正东 Python畅销书籍','6130ebb1f9e2c229.jpg');
```

3. 订单表

该表主要用于存储订单的相关信息，包括订单日期、订单付款状态和订单总价，其表结构如表 4-3 所示。

表 4-3 订单表的表结构

字 段 名	数据类型	长 度	主 键	外 键	备 注
id	varchar(20)	20	YES	NO	订单 id
order_date	varchar(20)	20	NO	NO	订单日期
status	integer	—	NO	NO	订单付款状态
total	float	—	NO	NO	订单总价

4. 详细订单表

该表主要用于存储详细订单的相关信息，包括图书 id、订单 id、图书数量和订单价格，其表结构如表 4-4 所示。

表 4-4 详细订单表的表结构

字 段 名	数据类型	长 度	主 键	外 键	备 注
id	integer	—	YES	NO	详细订单 id
goodsid	integer	—	NO	YES	图书 id
orderid	integer	—	NO	YES	订单 id
quantity	integer	—	NO	NO	图书数量
sub_total	float	—	NO	NO	订单价格

4.3 程序目录结构

该项目的目录结构如图 4-8 所示。

其中,文件夹 db 用于存放与数据库相关的文件;文件夹 static 用于存放静态资源;文件夹 templates 用于存放模板文件;文件 app.py 为项目的主程序;文件 config.py 用于编写项目的相关配置;文件 exts.py 用于定义数据库对象,以防止发生循环引用;文件 forms.py 用于编写表单验证;文件 manage.py 用于编写数据库初始化的相关命令;文件 models.py 用于编写数据模型。

图 4-8 目录结构

4.4 程序编写

在创建完项目的数据库及明确了项目的目录结构之后,开始进行程序编写。

(1) 使用 PyCharm(Professional)创建 Flask 项目 BookStore,如图 4-9 所示。

(2) 在项目根目录下创建文件夹 db,并将文件 store-dataload.sql 和 store-schema.sql 存入该文件夹中。

(3) 在项目根目录下创建文件 config.py,用于编写项目的相关配置,代码如下:

图 4-9 创建 Flask 项目

```python
# 资源包\BookStore\config.py
import os
DEBUG = True
SQLALCHEMY_DATABASE_URI = "sqlite://db/database.db"
SQLALCHEMY_TRACK_MODIFICATIONS = False
SQLALCHEMY_ECHO = True
SECRET_KEY = os.urandom(24)
```

(4) 完善 app.py 文件,导入配置文件 config.py,代码如下:

```python
# 资源包\BookStore\app.py
from flask import Flask
import config
app = Flask(__name__)
app.config.from_object(config)
if __name__ == '__main__':
    app.run()
```

(5) 在项目的根目录下的文件夹 db 中创建文件 dbhelper.py,用于编写创建数据库和插入数据的代码,代码如下:

```python
# 资源包\BookStore\db\dbhelper.py
import sqlite3
DB_FILES = './db/database.db'
def create_tables():
    f_name = 'db/store-schema.sql'
```

```python
        with open(f_name, 'r', encoding = 'utf-8') as f:
            sql = f.read()
            conn = sqlite3.connect(DB_FILES)
            try:
                conn.executescript(sql)
                print('数据库初始化成功')
            except Exception as e:
                print('数据库初始化失败')
                print(e)
            finally:
                conn.close()
def load_data():
    f_name = './db/store-dataload.sql'
    with open(f_name, 'r', encoding = 'utf-8') as f:
        sql = f.read()
        conn = sqlite3.connect(DB_FILES)
        try:
            conn.executescript(sql)
            print('数据库插入成功')
        except Exception as e:
            print('数据库插入失败')
            print(e)
        finally:
            conn.close()
```

（6）在项目根目录下创建文件 manage.py，用于编写数据库初始化的相关命令，代码如下：

```python
#资源包\BookStore\manage.py
from flask_script import Manager
from app import app
from db import dbhelper
manager = Manager(app)
@manager.command
def create_tables():
    dbhelper.create_tables()
@manager.command
def load_data():
    dbhelper.load_data()
if __name__ == '__main__':
    manager.run()
```

（7）打开命令提示符窗口，进入项目所在的根目录中，激活虚拟环境，并输入命令 python manage.py create_tables，即可完成数据库的创建，如图 4-10 所示。

此时，在文件夹 db 中创建了一个新的数据库文件 database.db，如图 4-11 所示。

图 4-10　创建数据库

图 4-11　目录 db 的结构

最后，在命令提示符中输入命令 python manage.py load_data，即可将数据导入数据库中，如图 4-12 所示。

（8）将文件夹 css、goods_images 和 images 存放至项目根目录下的文件夹 static 中，如图 4-13 所示。

图 4-12　向数据库中插入数据　　　图 4-13　static 目录结构

（9）在项目根目录下创建文件 models.py，用于编写数据模型，代码如下：

```python
# 资源包\BookStore\models.py
from app import db
# 客户表
class Customer(db.Model):
    __tablename__ = 'customers'
    id = db.Column('id', db.String(20), primary_key=True)
    name = db.Column('name', db.String(50), nullable=False)
    password = db.Column('password', db.String(20), nullable=False)
    address = db.Column('address', db.String(100))
    phone = db.Column('phone', db.String(20))
    birthday = db.Column('birthday', db.String(20))
# 商品表
class Goods(db.Model):
    __tablename__ = 'goods'
    id = db.Column('id', db.Integer, primary_key=True)
    name = db.Column('name', db.String(100), nullable=False)
    author = db.Column('author', db.String(30))
    press = db.Column('press', db.String(200))
    isbn = db.Column('isbn', db.String(30))
    edition = db.Column('edition', db.String(30))
    packaging = db.Column('packaging', db.String(30))
    format = db.Column('format', db.String(30))
    publication_time = db.Column('publication_time', db.String(30))
    paper = db.Column('paper', db.String(30))
    description = db.Column('description', db.String(200))
    price = db.Column('price', db.String(30))
    image = db.Column('image', db.String(100))
    orderLineItems = db.relationship('OrderLineItem')
# 订单表
class Orders(db.Model):
    __tablename__ = 'orders'
    id = db.Column('id', db.String(20), primary_key=True)
    orderdate = db.Column('order_date', db.String(20))
    # 1 表示待付款；0 表示已付款
    status = db.Column('status', db.Integer)
    total = db.Column('total', db.Float)
    orderLineItems = db.relationship('OrderLineItem')
# 详细订单表
class OrderLineItem(db.Model):
    __tablename__ = 'orderLineItems'
    id = db.Column('id', db.Integer, primary_key=True)
    quantity = db.Column('quantity', db.Integer)
    subtotal = db.Column('sub_total', db.Float)
```

```python
goodsid = db.Column('goodsid', db.ForeignKey('goods.id'))
orderid = db.Column('orderid', db.ForeignKey('orders.id'))
orders = db.relationship('Orders', backref = 'OrderLineItem')
goods = db.relationship('Goods', backref = 'OrderLineItem')
```

（10）在项目根目录下创建文件 exts.py，用于定义数据库对象，代码如下：

```python
# 资源包\BookStore\exts.py
from flask_sqlalchemy import SQLAlchemy
db = SQLAlchemy()
```

（11）在项目根目录下的文件夹 templates 中创建父模板 base_header.html，用于显示带有统一标题的页面，代码如下：

```html
# 资源包\BookStore\templates\base_header.html
<!doctype html>
<html>
    <head>
        <meta charset = "utf-8">
        <title>{% block title %}{% endblock %}</title>
        <link rel = "stylesheet" type = "text/css" href = "{{ url_for('static', filename = '../static/css/public.css') }}">
    </head>
    <body>
    <div class = "header">网上图书商城</div>
    <hr width = "100%"/>
    {% with messages = get_flashed_messages() %}
        {% if messages %}
            <ul>
                {% for message in messages %}
                    <li class = "success">{{ message }}</li>
                {% endfor %}
            </ul>
        {% endif %}
    {% endwith %}
    {% block body %}
    {% endblock %}
    <div class = "footer">
        <hr width = "100%"/>
        Copyright @老夏学院 2016-2024. All Rights Reserved
    </div>
    </body>
</html>
```

（12）在项目根目录下的文件夹 templates 中创建父模板 base_title.html，用于显示带有自定义标题的页面，代码如下：

```html
# 资源包\BookStore\templates\base_title.html
<!doctype html>
<html>
    <head>
        <meta charset = "utf-8">
        <title>{% block title %}{% endblock %}</title>
```

```html
    <link rel="stylesheet" type="text/css" href="{{ url_for('static', filename='../static/css/public.css') }}">
    </head>
    <body>
    {% with messages = get_flashed_messages() %}
        {% if messages %}
            <ul>
                {% for message in messages %}
                    <li class="success">{{ message }}</li>
                {% endfor %}
            </ul>
        {% endif %}
    {% endwith %}
    {% block body %}
    {% endblock %}
    <div class="footer">
        <hr width="100%"/>
        Copyright @老夏学院 2016-2024. All Rights Reserved
    </div>
    </body>
</html>
```

(13) 在项目根目录下的文件夹 templates 中创建模板文件 login.html,用于显示客户登录页面,代码如下:

```html
#资源包\BookStore\templates\login.html
{% extends "base_header.html" %}
{% block title %}客户登录{% endblock %}
{% block body %}
    <!-- 显示登录错误信息 -->
    <ul>
        {% for field, errors in form.errors.items() %}
            {% for message in errors %}
                <li class="error">{{ message }}</li>
            {% endfor %}
        {% endfor %}
    </ul>
    <form action="/login" method="post">
        <input type="hidden" name="csrf_token" value="{{ csrf_token() }}">
        <table width="100%" align="center">
            <tr height="40">
                <td colspan="2" align="center"><strong>请您登录</strong></td>
            </tr>
            <tr height="40">
                <td width="50%" align="right"><img src="{{ url_for('static', filename='images/3.jpg') }}"
align="absmiddle"/>  客户账号:
                </td>
                <td>{{ form.userid }}</td>
            </tr>
            <tr height="40">
                <td width="50%" align="right"><img src="{{ url_for('static', filename='images/2.jpg') }}"
```

```
align = "absmiddle"/>   客户密码:
            </td>
            <td>{{ form.password }}</td>
        </tr>
        <tr height = "40">
            <td align = "right">  </td>
            <td>< input type = "image" src = "{{ url_for('static', filename = 'images/login_button.jpg') }}"
                    onclick = "document.forms[0].fn.value = 'login'"/>
                     < a href = "/reg">< img
                    src = "{{ url_for('static', filename = 'images/reg_button.jpg') }}"
border = "0"/></a></td>
        </tr>
    </table>
</form>
{% endblock %}
```

(14) 在项目根目录下的文件夹 templates 中创建模板文件 customer_reg.html,用于显示客户注册页面,代码如下:

```
#资源包\BookStore\templates\customer_reg.html
{% extends "base_title.html" %}
{% block title %}用户注册{% endblock %}
{% block body %}
    < style type = "text/css">
        table {
            border - collapse: collapse;
        }
        .boder {
            border: 1px solid #5B96D0;
        }
        .col1 {
            background - color: #A6D2FF;
            text - align: right;
            padding - right: 10px;
            border: 1px solid #5B96D0;
            line - height: 50px;
        }
        .col2 {
            padding - left: 10px;
            border: 1px solid #5B96D0;
            line - height: 50px;
        }
        .textfield {
            height: 20px;
            width: 200px;
            border: 1px solid #999999;
            text - align: left;
            font - size: medium;
            line - height: 50px;
        }
    </style>
    < div >< img src = "{{ url_for('static', filename = '../static/images/reg.jpg') }}" align = "absmiddle"/></div>
```

```html
<br>
<hr width="100%"/>
<!-- 显示登录错误信息 -->
<ul>
    {% for field, errors in form.errors.items() %}
        {% for message in errors %}
            <li class="error">{{ message }}</li>
        {% endfor %}
    {% endfor %}
</ul>
<div class="text3" align="center">请填写下列信息</div>
<br>
<form action="/reg/" method="POST">
    <input type="hidden" name="csrf_token" value="{{ csrf_token() }}">
    <table width="60%" border="0" align="center" class="boder">
        <tr>
            <td width="35%" height="27" class="col1">客户账号:</td>
            <td class="col2">{{ form.userid(class='input') }} *</td>
        </tr>
        <tr>
            <td height="27" class="col1">客户姓名:</td>
            <td class="col2">{{ form.name(class='input') }} *</td>
        </tr>
        <tr>
            <td height="27" class="col1">客户密码:</td>
            <td class="col2">{{ form.password(class='input') }} *</td>
        </tr>
        <tr>
            <td height="27" class="col1">再次输入密码:</td>
            <td class="col2">{{ form.password2(class='input') }} *</td>
        </tr>
        <tr>
            <td height="27" class="col1">出生日期:</td>
            <td class="col2">{{ form.birthday(class='input') }} *
                格式(YYYY-MM-DD)
            </td>
        </tr>
        <tr>
            <td height="27" class="col1">通信地址:</td>
            <td class="col2">{{ form.address(class='input') }}</td>
        </tr>
        <tr>
            <td height="27" class="col1">电话号码:</td>
            <td class="col2">{{ form.phone(class='input') }}</td>
        </tr>
    </table>
    <br>
    <div align="center">
        <input type="image" src="{{ url_for('static', filename='../static/images/submit_button.jpg') }}"/>
    </div>
</form>
{% endblock %}
```

(15) 在项目根目录下的文件夹 templates 中创建模板文件 customer_reg_success.

html,用于显示客户注册成功后的页面,代码如下:

```
#资源包\BookStore\templates\customer_reg_success.html
{% extends "base_header.html" %}
{% block title %}注册成功{% endblock %}
{% block body %}
    <style type="text/css">
        a:link {
            font-size: 18px;
            color: #DB8400;
            text-decoration: none;
            font-weight: bolder;
        }
        a:visited {
            font-size: 18px;
            color: #DB8400;
            text-decoration: none;
            font-weight: bolder;
        }
        a:hover {
            font-size: 18px;
            color: #DB8400;
            text-decoration: underline;
            font-weight: bolder;
        }
    </style>
    <div align="center">
        <p class="text7">恭喜您注册成功!</p>
        <p>
            <a href="/login">返回登录页面</a>
        </p>
    </div>
{% endblock %}
```

(16)在项目根目录下创建文件 form.py,用于编写表单验证代码,代码如下:

```
#资源包\BookStore\forms.py
from flask_wtf import FlaskForm
from wtforms import StringField, PasswordField
from wtforms.validators import Length, EqualTo, InputRequired, Regexp
class LoginForm(FlaskForm):
    '''客户登录页面中的表单'''
    userid = StringField('客户账号: ', validators=[Length(min=3, max=10, message="客户账号长度必须在3到10位"), InputRequired('客户账号必须输入')])
    password = PasswordField('客户密码: ', validators=[InputRequired('客户姓名必须输入')])
class CustomerRegForm(FlaskForm):
    '''客户注册页面中的表单'''
    userid = StringField(label='客户账号', validators=[Length(min=3, max=10, message="客户账号长度必须在3到10位"), InputRequired('客户账号必须输入')])
    name = StringField(label='客户姓名: ', validators=[Length(min=3, max=10, message="客户姓名长度必须在3到10位"), InputRequired('客户姓名必须输入')])
    password = PasswordField(label='客户密码: ', validators=[Length(min=6, max=10, message="客户密码长度必须在6到10位"), InputRequired('客户密码必须输入')])
    password2 = PasswordField(label='再次输入密码: ', validators=[Length(min=6, max=10, message="客户密码长度必须在6到10位"), EqualTo("password", message='两次输入的密码不一致')])
```

```python
# 验证日期的正则表达式 YYYY-MM-DD YY-MM-DD
reg_date = r'^((((19|20)(([02468][048])|([13579][26]))-02-29))|((20[0-9][0-9])|(19[0-9][0-9]))-(((0[1-9])|(1[0-2]))-((0[1-9])|(1\d)|(2[0-8]))|((0[13578]|(1[02]))-31)|(((0[1,3-9])|(1[0-2]))-(29|30)))))$'
birthday = StringField(label='出生日期：', validators=[Regexp(reg_date, message='输入的日期无效')])
address = StringField(label='通信地址：')
phone = StringField(label='电话号码：')
```

（17）在项目根目录下的文件夹 templates 中创建模板文件 main.html，用于显示登录成功后的主页面，代码如下：

```html
# 资源包\BookStore\templates\main.html
{% extends "base_header.html" %}
{% block title %}主页面{% endblock %}
{% block body %}
    <style type="text/css">
        a:link {
            font-size: 18px;
            color: #DB8400;
            text-decoration: none;
            font-weight: bolder;
        }
        a:visited {
            font-size: 18px;
            color: #DB8400;
            text-decoration: none;
            font-weight: bolder;
        }
        a:hover {
            font-size: 18px;
            color: #DB8400;
            text-decoration: underline;
            font-weight: bolder;
        }
    </style>
    <div>
        <p class="text1"><img src="{{ url_for('static', filename='images/4.jpg') }}" align="absmiddle"/><a
                    href="/list">商品列表</a></p>
        <p class="text2">您可以从产品列表中浏览感兴趣的产品进行购买</p>
    </div>
    <hr width="100%"/>
    <div>
        <p class="text1"><img src="{{ url_for('static', filename='images/mycar1.jpg') }}" align="absmiddle"/><a
                    href="/cart">购物车</a></p>
        <p class="text2">您可以把感兴趣的商品暂时放在购物车中</p>
    </div>
{% endblock %}
```

（18）在项目根目录下的文件夹 templates 中创建父模板文件 goods_header.html，用于显示购物车、我的账号和商品列表等快捷链接，代码如下：

```
# 资源包\BookStore\templates\goods_header.html
<td width = "734" align = "right">
    <img src = "{{ url_for('static', filename = 'images/mycar1.jpg') }}" align = "absmiddle"/>
    <a href = "/cart">购物车</a>|
    <a href = "/user">我的账号</a> |
    <a href = "/list">商品列表</a>
</td>
```

（19）在项目根目录下的文件夹 templates 中创建模板文件 goods_list.html，用于显示商品列表页面，代码如下：

```
# 资源包\BookStore\templates\goods_list.html
{% extends "base_title.html" %}
{% block title %}商品列表{% endblock %}
{% block body %}
    <style type = "text/css">
        table {
            border-collapse: collapse;
        }
        .col1 {
            padding-top: 5px;
            border-top: 1px dashed #666666;
            text-indent: 40px;
        }
        .col2 {
            padding-top: 5px;
            border-top: 1px dashed #666666;
            text-align: right;
        }
        .col3 {
            padding-top: 5px;
            border-top: 1px dashed #666666;
            text-align: center;
        }
    </style>
    <table width = "100%" border = "0" align = "center">
        <tr>
            <td width = "616"><img src = "{{ url_for('static', filename = 'images/list.jpg') }}" align = "absmiddle"/></td>
            {% include 'goods_header.html' %}
        </tr>
    </table>
    <hr width = "100%"/>
    <div class = "text3" align = "center">请从商品列表中选择您喜爱的商品</div>
    <br>
    <table width = "100%" border = "0" align = "center">
        <tr bgcolor = "#b4c8ed">
            <th>商品名称</th>
            <th width = "5%">商品价格</th>
            <th width = "15%">添加到购物车</th>
        </tr>
        {% for goods in list %}
            <tr bgcolor = {{ loop.cycle('#ffffff', '#edf8ff') }}>
```

```html
            <td class = "col1"><a href = "/detail?id = {{ goods.id }}">{{ goods.description }}</a></td>
            <td class = "col2">¥{{ goods.price }}</td>
            <td class = "col3"><a
                href = "/add?id = {{ goods.id }}&name = {{ goods.name }}&price = {{ goods.price }}">添加到购物车</a></td>
        </tr>
        {% endfor %}
    </table>
{% endblock %}
```

（20）在项目根目录下的文件夹 templates 中创建模板文件 goods_detail.html，用于显示商品详细信息页面，代码如下：

```html
#资源包\BookStore\templates\goods_detail.html
{% extends "base_title.html" %}
{% block title %}商品详细{% endblock %}
{% block body %}
    <style type = "text/css">
        .title {
            font-size: 20px;
            color: #FF6600;
            font-style: italic;
        }
    </style>
    <table width = "100%" border = "0" align = "center">
        <tr>
            <td width = "616"><img src = "{{ url_for('static', filename = 'images/info.jpg') }}" align = "absmiddle"/></td>
            {% include 'goods_header.html' %}
        </tr>
    </table>
    <hr width = "100%"/>
    <div class = "text3" align = "center">{{ goods.name }}</div>
    <table width = "100%" border = "0" align = "center">
        <tr>
            <td width = "40%" align = "right">
                <div><img src = "{{ url_for('static', filename = 'goods_images/') }}{{ goods.image }}" width = "360px"
                        height = "360px"/></div>
                <br></td>
            <td>
                <div align = "center" class = "text4">一 口 价：<span class = "title">{{ goods.price }}元</span></div>
                <br>
                <table width = "80%" height = "200" border = "0">
                    <tbody>
                    <tr>
                        <td width = "25%" class = "text5">作者：</td>
                        <td width = "25%" class = "text6">{{ goods.author }}</td>
                        <td width = "25%" class = "text5">出版社：</td>
                        <td width = "25%" class = "text6">{{ goods.press }}</td>
                    </tr>
                    <tr>
```

```html
                    < td class = "text5"> ISBN: </td >
                    < td class = "text6">{{ goods.isbn }}</td >
                    < td class = "text5">版次: </td >
                    < td class = "text6">{{ goods.edition }}</td >
                </tr >
                < tr >
                    < td class = "text5">开本: </td >
                    < td class = "text6">{{ goods.format }}</td >
                    < td class = "text5">出版时间: </td >
                    < td class = "text6">{{ goods.publication_time }}</td >
                </tr >
                < tr >
                    < td class = "text5">用纸: </td >
                    < td class = "text6">{{ goods.paper }}</td >
                    < td class = "text5">包装: </td >
                    < td class = "text6">{{ goods.packaging }}</td >
                </tr >
            </tbody >
        </table >
        < br >
        < br >
        < div >< a href = "/add?id = {{ goods.id }}&name = {{ goods.name }}&price = {{ goods.price }}">
            < img src = "{{ url_for('static', filename = 'images/button.jpg') }}"/>
        </a >
        </div >
    </td >
</tr >
</table >
{% endblock %}
```

（21）在项目根目录下的文件夹 templates 中创建模板文件 cart.html，用于显示购物车页面，代码如下：

```html
#资源包\BookStore\templates\cart.html
{% extends "base_title.html" %}
{% block title %}购物车{% endblock %}
{% block body %}
    < style type = "text/css">
        table {
            border - collapse: collapse;
        }
        .threeboder {
            border: 1px solid #5B96D0;
        }
        .trow {
            border - right: 1px solid #5B96D0;
            border - bottom: 1px solid #5A96D6;
        }
        .theader {
            background - color: #A5D3FF;
            font - size: 14px;
            border - right: 1px solid #5B96D0;
            border - bottom: 1px solid #5A96D6;
```

```
        }
    </style>
    <script>
        function calc(rowid, quantityInput) {
            quantity = quantityInput.value;
            if (isNaN(quantity)) {
                alert("不是有效的数值!");
                quantityInput.value = 0;
                quantity = quantityInput.value
                quantityInput.focus();
            }
            var price_id = 'price_' + rowid;
            var price = parseFloat(document.getElementById(price_id).innerText);
            var subtotal_id = 'subtotal_' + rowid;
            subtotal1 = parseFloat(document.getElementById(subtotal_id).innerText);
            subtotal1 = subtotal1.toFixed(2);
            document.getElementById(subtotal_id).innerText = quantity * price;
            subtotal2 = parseFloat(document.getElementById(subtotal_id).innerText);
            total = parseFloat(document.getElementById('total').innerText);
            total = total - subtotal1 + subtotal2;
            total = total.toFixed(2);
            document.getElementById('total').innerText = total;
        }
    </script>
    <table width="100%" border="0" align="center">
        <tr>
            <td width="616"><img src="{{url_for('static', filename='images/mycar.jpg') }}"/></td>
            {% include 'goods_header.html' %}
        </tr>
    </table>
    <hr width="100%"/>
    <div class="text3" align="center">您选好的商品</div>
    <br>
    <form action="/submit_order" method="post">
        <table width="100%" border="0" align="center" class="threeboder">
            <tr bgcolor="#A5D3FF">
                <td height="50" align="center" class="theader">商品名称</td>
                <td width="8%" align="center" class="theader">数量</td>
                <td width="15%" align="center" class="theader">单价</td>
                <td width="15%" align="center" class="theader">小计</td>
            </tr>
            {% for item in list %}
                <tr>
                    <td height="50" align="left" class="trow">{{ item[1] }}</td>
                    <td align="center" class="trow">
                        <input name="quantity_{{item[0]}}" type="text" value="{{item[3]}}"
                               onblur="calc({{ item[0] }}, this)"/>
                    </td>
                    <td align="center" class="trow">&yen;<span id="price_{{item[0]}}">{{ item[2]}}</span></td>
                    <td align="center" class="trow">&yen;<span id="subtotal_{{item[0]}}">{{ item[4] }}</span>
                    </td>
                </tr>
```

```html
            {% endfor %}
            <tr>
                <td height="50" colspan="5" align="right">合计：&yen;<span id="total">{{ total }}</span>  </td>
            </tr>
        </table>
        <br>
        <div align="center">
            <input type="image" src="{{ url_for('static', filename='images/submit_order.jpg') }}"/>  
        </div>
    </form>
{% endblock %}
```

（22）在项目根目录下的文件夹 templates 中创建模板文件 order_finish.html，用于显示订单完成页面，代码如下：

```html
#资源包\BookStore\templates\order_finish.html
{% extends "base_header.html" %}
{% block title %}订单完成{% endblock %}
{% block body %}
    <style type="text/css">
        a:link {
            font-size: 18px;
            color: #DB8400;
            text-decoration: none;
            font-weight: bolder;
        }
        a:visited {
            font-size: 18px;
            color: #DB8400;
            text-decoration: none;
            font-weight: bolder;
        }
        a:hover {
            font-size: 18px;
            color: #DB8400;
            text-decoration: underline;
            font-weight: bolder;
        }
    </style>
    <div align="center">
        <p class="text7">谢谢您的购物！</p>
        <p class="text7">您的订单号是：{{ orderid }}</p>
        <p class="text7">您可以继续购物！</p>
        <p class="text7">
            <a href="/main">返回主页面</a>
        </p>
    </div>
{% endblock %}
```

（23）在项目根目录下的文件夹 templates 中创建模板文件 user.html，用于显示账户信息页面，代码如下：

```
#资源包\BookStore\templates\user.html
{% extends "base_title.html" %}
{% block title %}账户信息{% endblock %}
{% block body %}
    <style type="text/css">
        table {
            border-collapse: collapse;
        }
        .boder {
            border: 1px solid #5B96D0;
        }
        .col1 {
            background-color: #A6D2FF;
            text-align: right;
            padding-right: 10px;
            border: 1px solid #5B96D0;
            line-height: 50px;
        }
        .col2 {
            padding-left: 10px;
            border: 1px solid #5B96D0;
            line-height: 50px;
        }
        .textfield {
            height: 20px;
            width: 200px;
            border: 1px solid #999999;
            text-align: left;
            font-size: medium;
            line-height: 50px;
        }
    </style>
    <table width="100%" border="0" align="center">
        <tr>
            <td width="616"><img src="{{ url_for('static', filename='../static/images/user.jpg') }}" align="absmiddle"/></td>
            {% include 'goods_header.html' %}
        </tr>
    </table>
    <hr width="100%"/>
    <div class="text3" align="center">账户信息</div>
    <br>
    <form action="/reg" method="POST">
        <table width="60%" border="0" align="center" class="boder">
            <tr>
                <td width="35%" height="27" class="col1">客户账号：</td>
                <td class="col2">{{ user.id }}</td>
            </tr>
            <tr>
                <td height="27" class="col1">客户姓名：</td>
                <td class="col2">{{ user.id }}</td>
            </tr>
            <tr>
                <td height="27" class="col1">出生日期：</td>
                <td class="col2">{{ user.birthday }}</td>
            </tr>
```

```html
            <tr>
                <td height = "27" class = "col1">通信地址: </td>
                <td class = "col2">{{ user.address }}</td>
            </tr>
            <tr>
                <td height = "27" class = "col1">电话号码: </td>
                <td class = "col2">{{ user.phone }}</td>
            </tr>
        </table>
        <br>
    </form>
{% endblock %}
```

(24) 完善 app.py 文件,代码如下:

```python
# 资源包\BookStore\app.py
from flask import Flask, request, session, render_template, redirect, url_for, flash
from exts import db
from flask_wtf import CSRFProtect
from forms import CustomerRegForm, LoginForm
from models import Customer, Goods, Orders, OrderLineItem
import config
import random
import datetime
app = Flask(__name__)
app.config.from_object(config)
csrf = CSRFProtect()
csrf.init_app(app)
db.init_app(app)
# 客户注册
@app.route('/reg/', methods = ['GET', 'POST'])
def register():
    form = CustomerRegForm()
    if request.method == 'POST':
        if form.validate():
            new_customer = Customer()
            new_customer.id = form.userid.data
            new_customer.name = form.name.data
            new_customer.password = form.password.data
            new_customer.address = form.address.data
            new_customer.birthday = form.birthday.data
            new_customer.phone = form.phone.data
            db.session.add(new_customer)
            db.session.commit()
            print('注册成功')
            return render_template('customer_reg_success.html', form = form)
    return render_template('customer_reg.html', form = form)
# 客户登录
@app.route('/')
@app.route('/login/', methods = ['GET', 'POST'])
def login():
    form = LoginForm()
    if request.method == 'POST':
        if form.validate():
            c = db.session.query(Customer).filter_by(id = form.userid.data).first()
            if c is not None and c.password == form.password.data:
                print('登录成功')
                customer = {}
```

```python
                customer['id'] = c.id
                customer['name'] = c.name
                customer['password'] = c.password
                customer['address'] = c.address
                customer['phone'] = c.phone
                customer['birthday'] = c.birthday
                session['customer'] = customer
                return redirect(url_for('main'))
            else:
                flash('您输入的客户账号和密码错误.')
                return render_template('login.html', form = form)
    return render_template('login.html', form = form)
#登录成功后的主页面
@app.route('/main/')
def main():
    if 'customer' not in session.keys():
        flash('您还没有登录,请登录.')
        return redirect(url_for('login'))
    return render_template('main.html')
#商品列表
@app.route('/list/')
def show_goods_list():
    if 'customer' not in session.keys():
        flash('您还没有登录,请登录.')
        return redirect(url_for('login'))
    goodslist = db.session.query(Goods).all()
    return render_template('goods_list.html', list = goodslist)
#商品详细
@app.route('/detail/')
def show_goods_detail():
    if 'customer' not in session.keys():
        flash('您还没有登录,请登录.')
        return redirect(url_for('login'))
    goodsid = request.args['id']
    goods = db.session.query(Goods).filter_by(id = goodsid).first()
    return render_template('goods_detail.html', goods = goods)
#添加购物车
@app.route('/add/')
def add_cart():
    if 'customer' not in session.keys():
        flash('您还没有登录,请登录.')
        return redirect(url_for('login'))
    goodsid = int(request.args['id'])
    goodsname = request.args['name']
    goodsprice = float(request.args['price'])
    if 'cart' not in session.keys():
        session['cart'] = []
    cart = session['cart']
    flag = 0
    for item in cart:
        if item[0] == goodsid:
            item[3] += 1
            flag = 1
            break
    if flag == 0:
        cart.append([goodsid, goodsname, goodsprice, 1])
    session['cart'] = cart
    print(cart)
    flash('已经添加商品【' + goodsname + '】到购物车')
```

```python
        return redirect(url_for('show_goods_list'))
#查看购物车
@app.route('/cart/')
def show_cart():
    if 'customer' not in session.keys():
        flash('您还没有登录,请登录.')
        return redirect(url_for('login'))
    if 'cart' not in session.keys():
        return render_template('cart.html', list = [], total = 0.0)
    cart = session['cart']
    list = []
    total = 0.0
    for item in cart:
        subtotal = item[2] * item[3]
        total += subtotal
        new_item = (item[0], item[1], item[2], item[3], subtotal)
        list.append(new_item)
    return render_template('cart.html', list = list, total = total)
#提交订单
@app.route('/submit_order/', methods = ['POST'])
def submit_order():
    orders = Orders()
    n = random.randint(0, 9)
    d = datetime.datetime.today()
    orderid = str(int(d.timestamp() * 1e6)) + str(n)
    orders.id = orderid
    orders.orderdate = d.strftime('%Y-%m-%d %H:%M:%S')
    orders.status = 1
    db.session.add(orders)
    cart = session['cart']
    total = 0.0
    for item in cart:
        quantity = request.form['quantity_' + str(item[0])]
        try:
            quantity = int(quantity)
        except:
            quantity = 0
        subtotal = item[2] * quantity
        total += subtotal
        order_line_item = OrderLineItem()
        order_line_item.quantity = quantity
        order_line_item.goodsid = item[0]
        order_line_item.orderid = orderid
        order_line_item.subtotal = subtotal
        db.session.add(order_line_item)
    orders.total = total
    db.session.commit()
    session.pop('cart', None)
    return render_template('order_finish.html', orderid = orderid)
#客户账户
@app.route('/user/')
def show_user():
    if 'customer' not in session.keys():
        flash('您还没有登录,请登录.')
        return redirect(url_for('login'))
    name = session['customer']['name']
    user = db.session.query(Customer).filter_by(name = name).first()
    return render_template('user.html', user = user)
```

第 5 章 Django

5.1 Django 简介

Django 是一个由 Python 编写的开放源代码的 Web 应用框架,用于快速开发可维护和可扩展的 Web 应用程序。

通过 Django,开发人员只需很少的代码,就可以轻松地完成一个正式网站所需要的大部分内容,并可以进一步开发出全功能的 Web 服务。

Django 是一个遵循 MVC 设计模式的框架,但由于其控制器中接收用户输入的部分是由框架自行处理的,所以 Django 更关注的是模型(Model)、模板(Template)和视图(Views),即 MTV 模式。

Django 的 MTV 模式本质上和 MVC 模式是一致的,均是为了各组件之间保持松耦合关系,只是在定义上有些许不同,其各部分的职责如下:

M 表示模型(Model),即数据存取层,主要用于处理与数据相关的所有事务,例如,如何存取、如何验证有效性、包含哪些行为及数据之间的关系等。

T 表示模板(Template),即表现层,主要用于处理与表现相关的业务,例如,如何在页面或其他类型文档中进行显示。

V 表示视图(View),即业务逻辑层,主要用于存取数据模型及调取恰当模板的相关逻辑,是数据模型与模板的桥梁。

5.2 安装 Django

由于 Django 属于 Python 的第三方库,所以需要进行安装,打开命令提示符窗口,输入命令 pip install django 即可。

5.3 第 1 个 Django 项目

创建 Django 项目可以通过两种方式,即使用 PyCharm(Professional)和使用命令提示符。

1. 使用 PyCharm(Professional)

1) 创建项目

通过 PyCharm(Professional)可以快速地创建一个 Django 项目,具体步骤如下:

（1）打开 PyCharm（Professional），选择 File→New Project，如图 5-1 所示。

图 5-1　创建 Django 项目的第 1 步

（2）在当前界面的左侧栏中选择 Django，然后在右侧栏中单击 Previously configured interpreter，并在其中选择虚拟环境 django_env，最后单击 Create 按钮，即可创建一个名为 djangoProject 的项目，如图 5-2 所示。

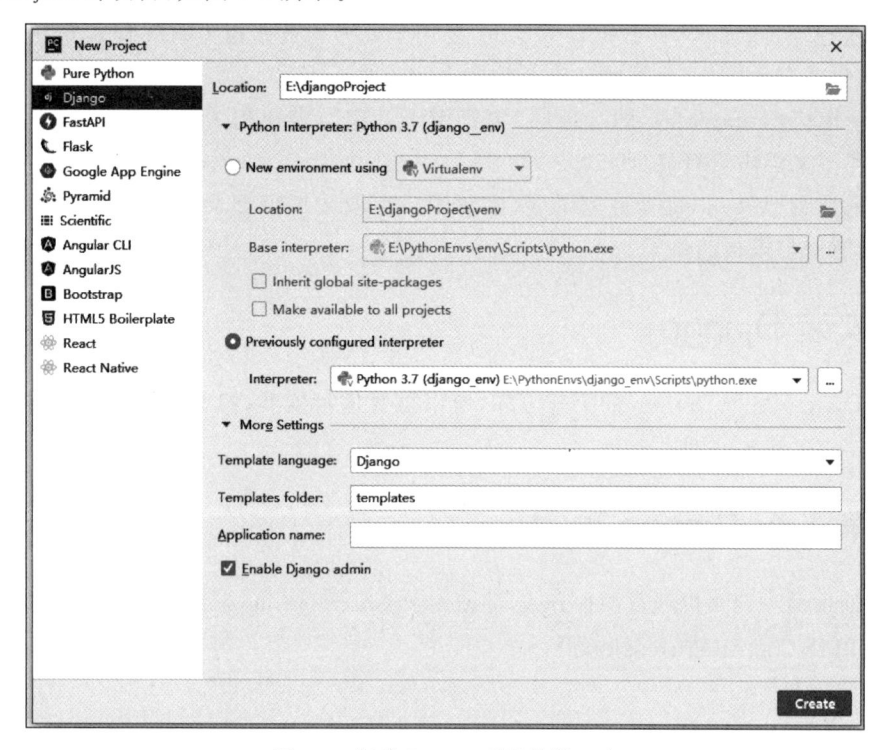

图 5-2　创建 Django 项目的第 2 步

此时,使用 PyCharm(Professional)创建的项目 djangoProject 的结构如图 5-3 所示。

其中,文件夹 djangoProject 表示项目中的应用,其与当前的项目名称相同;文件 manage.py 主要包含项目管理的子命令;文件夹 templates 主要用于存放模板文件;文件 asgi.py 为异步服务器网关接口的配置文件;文件 settings.py 为项目的配置文件;文件 wsgi.py 为 Web 服务网关的配置文件;文件 urls.py 为项目主路由的配置文件。

图 5-3 项目 djangoProject 的结构

这里重点讲解项目和应用的关系。应用是 Django 项目的组成部分,一个应用代表项目中的一个模块,所有 URL 请求的响应都是由应用来处理的,例如豆瓣网中有图书、电影、音乐和同城等模块,而这些模块在 Django 项目中就是应用,图书、电影、音乐和同城等应用共同组成了豆瓣网这个项目,因此一个项目中可以包含多个应用,并且一个应用也可以在多个项目中使用。

运行上述项目,PyCharm 中的显示结果如图 5-4 所示。

```
E:\PythonEnvs\django_env\Scripts\python.exe E:/djangoProject/manage.py runserver 8000
Performing system checks...

System check identified no issues (0 silenced).

You have 15 unapplied migration(s). Your project may not work properly until you apply the migrations for
 app(s): admin, auth, contenttypes, sessions.
Run 'python manage.py migrate' to apply them.
May 19, 2024 - 22:57:37
Django version 2.1.15, using settings 'djangoProject.settings'
Starting development server at http://127.0.0.1:8000/
Quit the server with CTRL-BREAK.
```

图 5-4 PyCharm 中的显示内容

此时,打开浏览器并在网址栏输入 http://127.0.0.1:8000,其显示内容如图 5-5 所示。

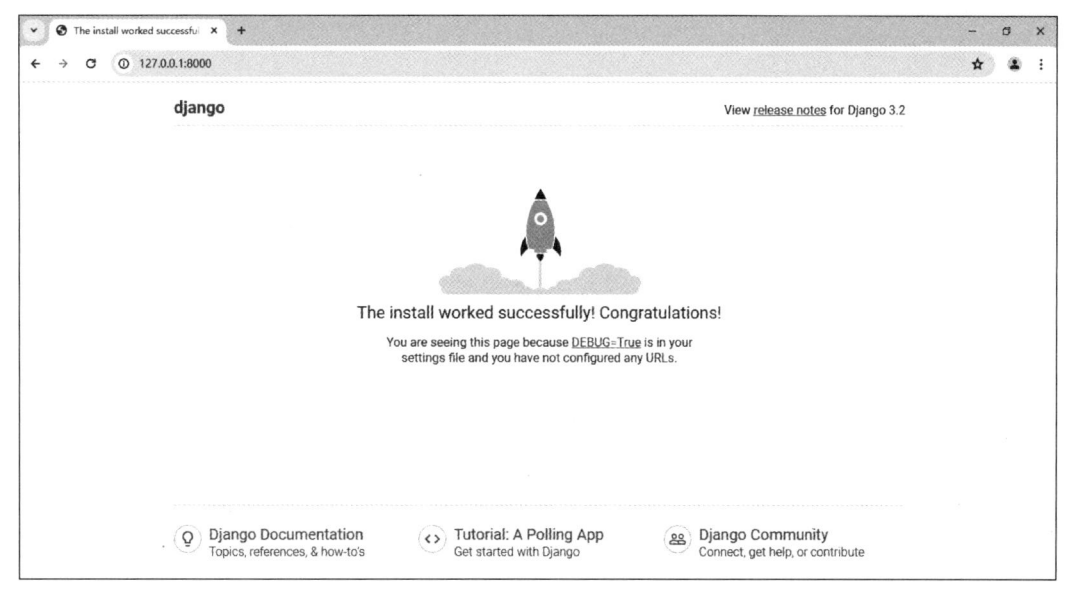

图 5-5 浏览器中的显示内容

2)修改端口号

通过单击 Edit Configurations,并在其显示的页面中修改 Port 所对应的值,即可完成端口号的修改,如图 5-6 所示。

图 5-6 修改端口号

此时,打开浏览器并在网址栏输入 http://127.0.0.1:8888,其显示内容如图 5-7 所示。

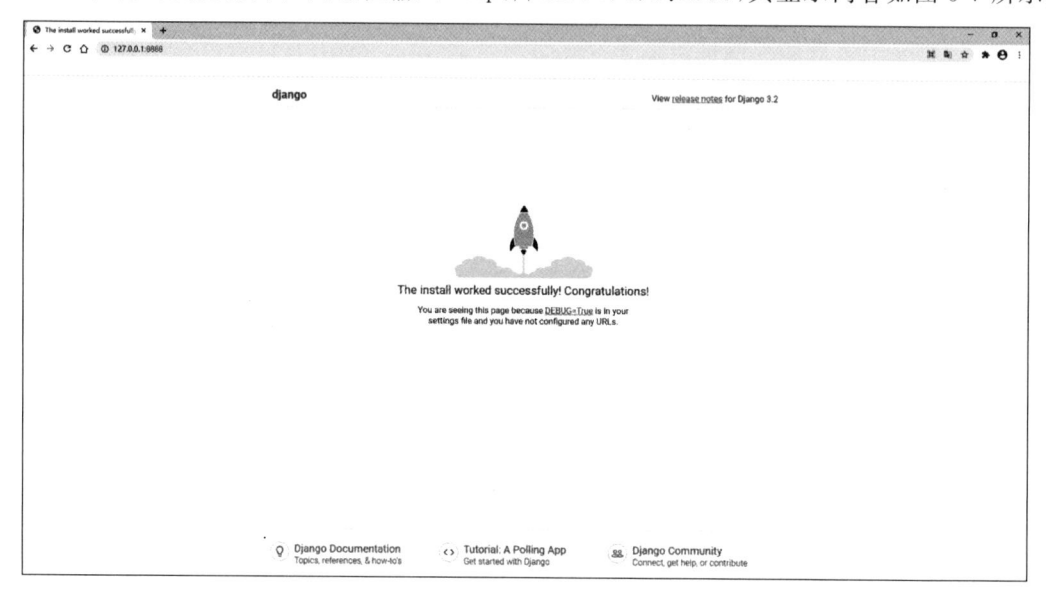

图 5-7 浏览器中的显示内容

3)外部访问

通过单击 Edit Configurations,并在其显示的页面中将 Host 所对应的值修改为 0.0.0.0

（如图 5-8 所示），然后将配置文件 settings. py 中的参数 ALLOWED_HOSTS 的值修改为本机的 IP，即可使服务器被外部访问。

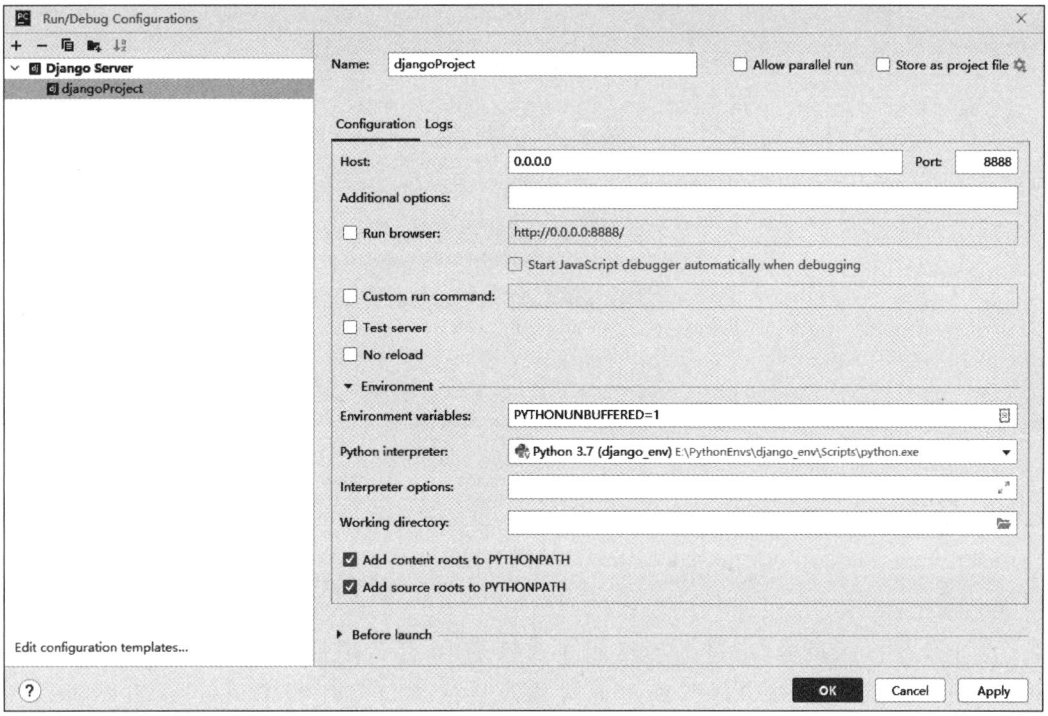

图 5-8　修改 Host

2. 使用命令提示符

1）创建项目

打开命令提示符窗口，进入需要创建项目的目录中，激活虚拟环境，并输入命令 django-admin startproject djangoProject，创建名为 djangoProject 的 Django 项目，如图 5-9 所示。

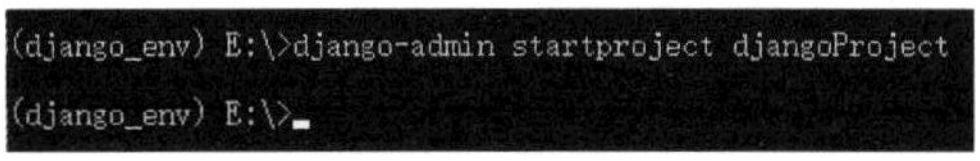

```
(django_env) E:\>django-admin startproject djangoProject

(django_env) E:\>_
```

图 5-9　创建项目

进入项目的根目录中，输入命令 python manage. py runserver 即可运行该项目，如图 5-10 所示。

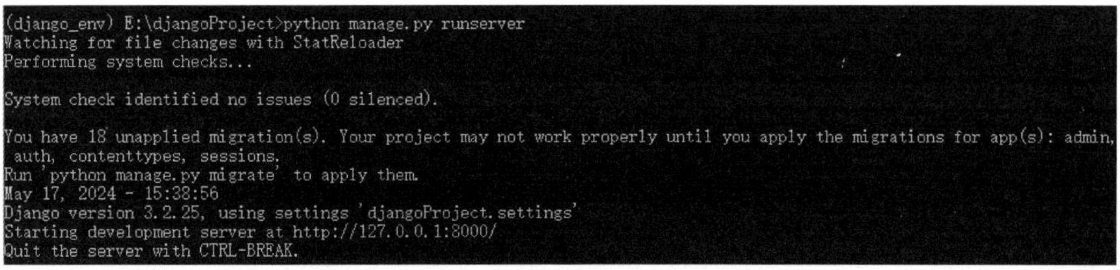

```
(django_env) E:\djangoProject>python manage.py runserver
Watching for file changes with StatReloader
Performing system checks...

System check identified no issues (0 silenced).

You have 18 unapplied migration(s). Your project may not work properly until you apply the migrations for app(s): admin,
 auth, contenttypes, sessions.
Run 'python manage.py migrate' to apply them.
May 17, 2024 - 15:38:56
Django version 3.2.25, using settings 'djangoProject.settings'
Starting development server at http://127.0.0.1:8000/
Quit the server with CTRL-BREAK.
```

图 5-10　运行项目

此时，打开浏览器并在网址栏输入 http://127.0.0.1:8000，其显示内容如图 5-11

所示。

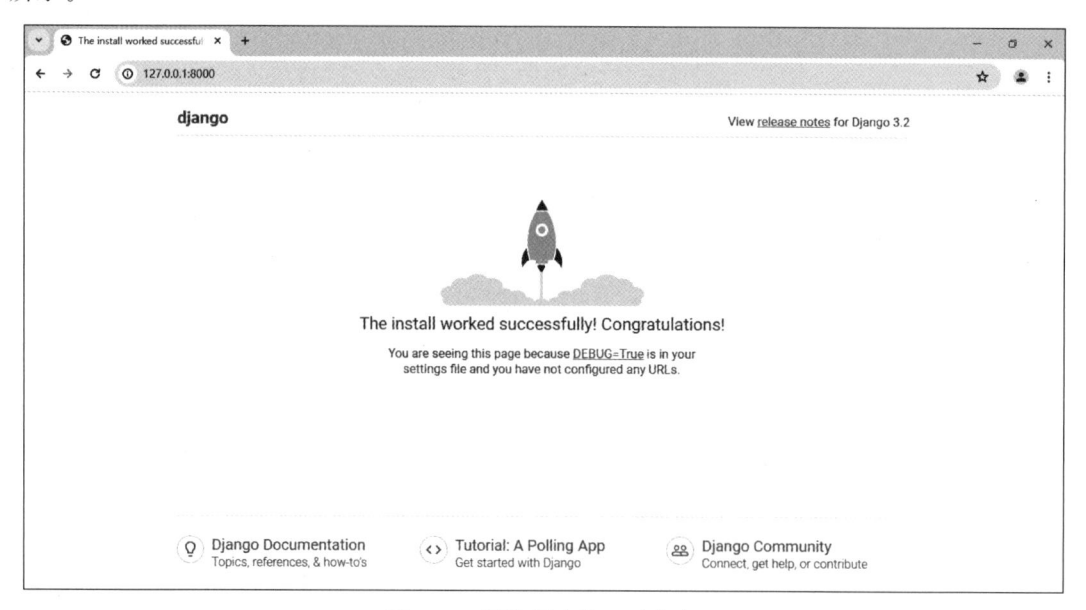

图 5-11　浏览器中的显示内容

2）修改端口号

打开命令提示符窗口,进入项目所在的目录中,激活虚拟环境,并输入命令"python manage.py runserver 端口号",即可通过修改后的端口号运行 Django 项目,如图 5-12 所示。

图 5-12　修改端口号

此时,打开浏览器并在网址栏输入 http://127.0.0.1:8888,其显示内容如图 5-13 所示。

3）外部访问

首先,将保存在项目根目录下的配置文件 settings.py 中的参数 ALLOWED_HOSTS 的值修改为本机的 IP,然后打开命令提示符窗口,进入项目所在的目录中,激活虚拟环境,并输入命令"python manage.py runserver 0.0.0.0:端口号",即可使服务器被外部访问,如图 5-14 所示。

此外,在 Django 中启用调试模式,需要将配置文件 settings.py 中的参数 DEBUG 的值设置为 True,并且需要确保参数 ALLOWED_HOSTS 的值包含本地开发主机。

需要注意的是,在生产环境中应将 DEBUG 设置为 False,否则将暴露关于当前应用和环境的敏感信息。

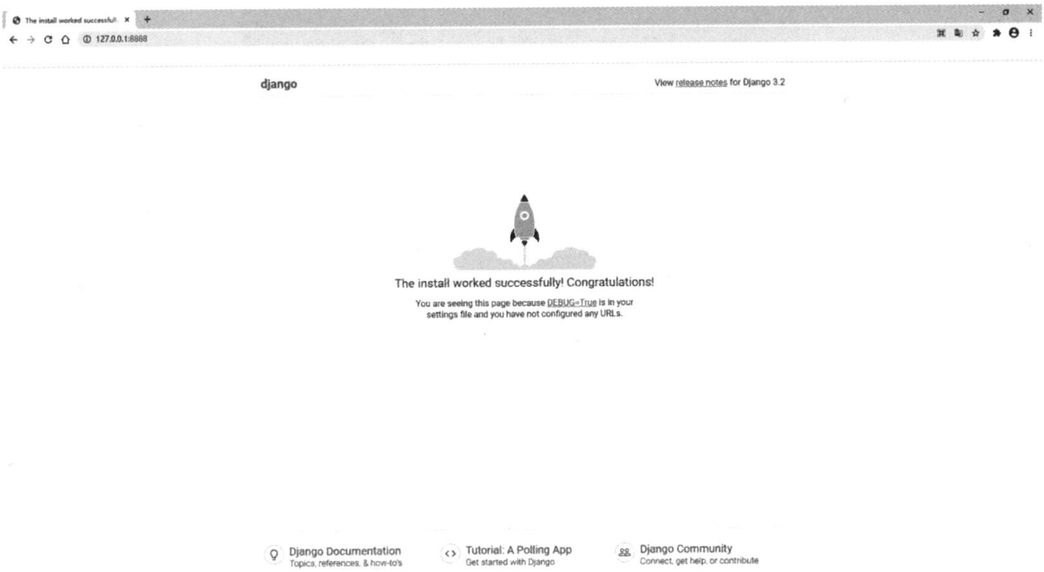

图 5-13　浏览器中的显示内容

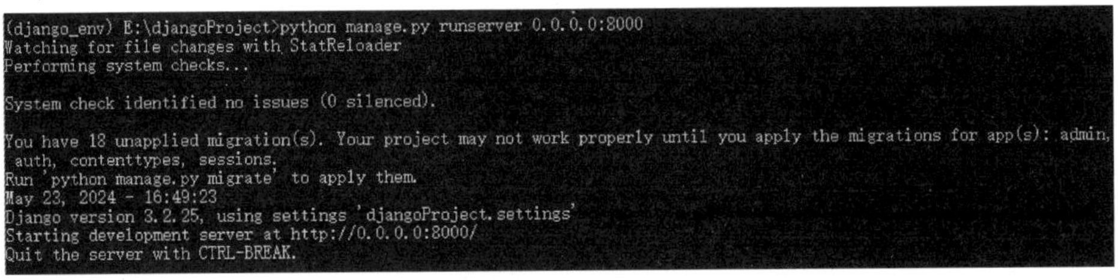

图 5-14　外部访问

5.4　路由

在 Django 中，用于处理请求 URL 和视图函数之间关系的程序称为路由。

5.4.1　视图函数

视图函数是一个简单的 Python 函数，其接受来自 Web 服务器的请求数据，并会将响应内容返给 Web 服务器。

一般情况下，视图函数均在应用中的文件 views.py 中进行编写。

视图函数中有两个重要的对象，即 HttpRequest 对象和 HttpResponse 对象。每个视图函数的第 1 个参数必须是 HttpRequest 对象，并且每个视图函数的返回结果都必须是一个 HttpResponse 对象。

下面通过一个示例演示一下如何创建视图函数。

（1）创建名为 djangoProject 的 Django 项目。

（2）打开命令提示符窗口，进入项目所在的根目录中，激活虚拟环境，并输入命令 python manage.py startapp book，创建名为 book 的应用，如图 5-15 所示。

```
E:\Python全栈开发\djangoProject>workon django_env
(django_env) E:\Python全栈开发\djangoProject>python manage.py startapp book
```

图 5-15　创建应用

(3) 打开应用 book 中的文件 views.py，代码如下：

```python
# 资源包\Code\chapter5\5.4\1\djangoProject\book\views.py
from django.http import HttpResponse
def book(request):
    return HttpResponse("这是书籍首页")
```

(4) 打开应用 djangoProject 中的文件 urls.py，编写 URL 映射（该部分内容将在后续章节为读者详细讲解），代码如下：

```python
# 资源包\Code\chapter5\5.4\1\djangoProject\djangoProject\urls.py
from django.urls import path
from django.http import HttpResponse
from book import views
def index(request):
    return HttpResponse('这是首页')
urlpatterns = [
    path('', index),
    path('book/', views.book)
]
```

(5) 运行上述项目，打开浏览器并在网址栏输入 http://127.0.0.1:8000/book/，其显示内容如图 5-16 所示。

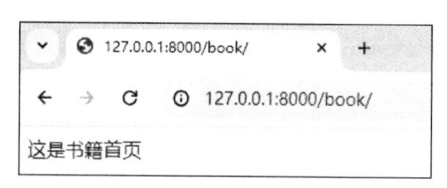

图 5-16　浏览器中的显示内容

5.4.2　URL 映射

在 Django 中，通过配置文件 settings.py 中的变量 ROOT_URLCONF 构建 URL 与视图函数的映射关系。

一般情况下，Django 项目中的路由通过 urls.py 文件进行配置，在该文件中定义了一个 urlpatterns 列表，而 URL 的定义就是在这个列表中完成的。

1. 普通路径

通过 django.urls 模块中的 path() 函数定义普通的 URL 路径，其语法格式如下：

```
path(route, view, name, kwargs)
```

其中，参数 route 表示 URL 路径，参数 view 表示视图函数，参数 name 表示 URL 命名，用于反向获取 URL，参数 kwargs 用于传递给视图函数额外的关键字参数。

2. 正则路径

通过 django.urls 模块中的 re_path() 函数定义匹配正则表达式的 URL 路径，其语法格式如下：

```
re_path(route, view, name, kwargs)
```

其中，参数 route 表示 URL 路径，该值可以为正则表达式，参数 view 表示视图函数，参数 name 表示 URL 命名，用于反向获取 URL，参数 kwargs 用于传递给视图函数额外的关键字参数。

下面通过一个示例演示一下如何定义 URL 映射。

(1) 创建名为 djangoProject 的 Django 项目。

(2) 打开命令提示符窗口，进入项目所在的根目录中，激活虚拟环境，并输入命令 python manage.py startapp book，创建名为 book 的应用，如图 5-17 所示。

图 5-17　创建应用

(3) 打开应用 book 中的文件 views.py，代码如下：

```
#资源包\Code\chapter5\5.4\2\djangoProject\book\views.py
from django.http import HttpResponse
def book(request):
    return HttpResponse("这是书籍首页")
def book_detail(request):
    return HttpResponse('这是书籍详细信息页面')
```

(4) 打开应用 djangoProject 中的文件 urls.py，代码如下：

```
#资源包\Code\chapter5\5.4\2\djangoProject\djangoProject\urls.py
from django.urls import path, re_path
from django.http import HttpResponse
from book import views
def index(request):
    return HttpResponse('这是首页')
urlpatterns = [
    path('', index),
    path('book/', views.book),
    re_path(r'book_detail/[0-9]{4}/', views.book_detail)
]
```

(5) 运行上述项目，打开浏览器并在网址栏输入 http://127.0.0.1:8000/book_detail/8888/，其显示内容如图 5-18 所示。

图 5-18　浏览器中的显示内容

5.4.3　HttpRequest 对象

HttpRequest 对象是 Django 中用于封装 HTTP 请求信息的类，即当用户发起一个

HTTP 请求时,Django 会创建一个 HttpRequest 对象,并将其传递给相应的视图函数。

HttpRequest 对象包含了请求的各种信息,例如 HTTP 方法、URL、查询参数和表单数据等,并且在视图函数中,可以通过 HttpRequest 对象的属性来获取请求的详细信息,其常用属性和方法如表 5-1 所示。

表 5-1 HttpRequest 对象的常用属性和方法

属 性	描 述
method	获取 HTTP 请求方法
path	获取请求的 URL 路径
GET	获取 GET 请求的参数,返回一个 QueryDict 对象
POST	获取 POST 请求的参数,返回一个 QueryDict 对象
FILES	获取所有的上传文件,返回一个 QueryDict 对象
COOKIES	获取所有的 Cookie 数据
session	获取或设置 Session 数据
META	获取所有的 HTTP 头部信息
user	获取当前用户信息

方 法	描 述
get_full_path()	获取请求的完整路径
get_raw_uri()	获取完整的原始 URL,包括域名和查询字符串
get_host()	获取请求的原始主机名和端口
is_secure()	判断请求是否采用 HTTPS 协议
is_ajax()	判断请求是否采用 AJAX 发送

5.4.4 QueryDict 对象

在 Django 中,QueryDict 是一个类似字典的对象,用于处理 HTTP 请求中的 GET、POST 或 FILES 等数据,其常用的方法如表 5-2 所示。

表 5-2 QueryDict 对象的常用方法

方 法	描 述
get()	获取指定键所对应的值。如果键不存在,则默认返回 None。此外,可以通过该方法的第 2 个参数 default 来指定默认值
getlist()	获取指定键所对应的所有值。如果键不存在,则返回空列表

5.4.5 HttpResponse 对象

在 Django 中,HttpResponse 对象用于表示 HTTP 响应,包含响应状态码、头部信息和主体内容等,其常用属性和方法如表 5-3 所示。

表 5-3 HttpResponse 对象的常用属性和方法

属 性	描 述
content	响应内容
status_code	响应状态码
content_type	响应的内容类型

方　　法	描　　述
set_cookie()	设置 Cookie
delete_cookie()	删除 Cookie
write()	用于将字符串数据写入响应中

5.4.6　JsonResponse 对象

在 Django 中，可以通过 JsonResponse 对象将字典直接转换为 JSON 响应。

5.4.7　重定向

重定向分为永久性重定向和暂时性重定向，在页面上体现的操作就是浏览器会从一个页面自动跳转到另一个页面，例如用户访问了一个需要权限的页面，但是该用户当前并没有登录，因此需要将该用户重定向至登录页面。

在 Django 中，可以通过 redirect() 函数实现重定向，其语法格式如下：

```
redirect(to, permanent)
```

其中，参数 to 表示重定向的 URL，参数 permanent 表示重定向的状态，默认为暂时性重定向，当其值为 True 时，表示永久性重定向。

5.4.8　动态路由

当请求 URL 需要动态变化时，需要使用动态路由，即需要给请求 URL 传递参数，可以通过将请求 URL 路径中的一部分标记为 < variable_name > 的方式添加参数。

这里需要注意两点，一是当请求 URL 中包含参数时，视图函数中必须传递该参数；二是当视图函数中的参数为默认参数时，请求 URL 中可以不用传递参数。

下面通过一个示例演示一下如何定义动态路由。

（1）创建名为 djangoProject 的 Django 项目。

（2）打开命令提示符窗口，进入项目所在的根目录中，激活虚拟环境，并输入命令 python manage.py startapp book，创建名为 book 的应用，如图 5-19 所示。

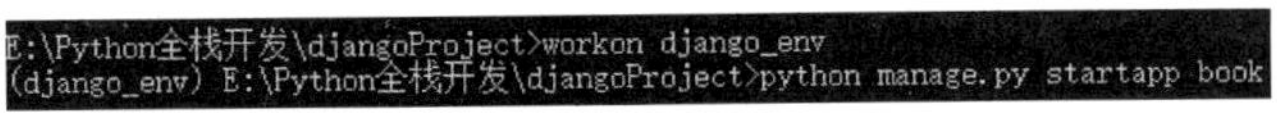

```
E:\Python全栈开发\djangoProject>workon django_env
(django_env) E:\Python全栈开发\djangoProject>python manage.py startapp book
```

图 5-19　创建应用

（3）打开应用 book 中的文件 views.py，代码如下：

```
#资源包\Code\chapter5\5.4\3\djangoProject\book\views.py
from django.http import HttpResponse
def book_info(request, book_info = "default"):
    return HttpResponse(f"书籍的信息为{book_info}")
def book_id(request, book_id):
    return HttpResponse(f"书籍的 id 为{book_id}")
```

（4）打开应用 djangoProject 中的文件 urls.py，代码如下：

```
#资源包\Code\chapter5\5.4\3\djangoProject\djangoProject\urls.py
from django.urls import path
from django.http import HttpResponse
from book import views
def index(request):
    return HttpResponse('这是首页')
urlpatterns = [
    path('', index),
    path('book_info/', views.book_info),
    path('book_id/<book_id>/', views.book_id),
]
```

（5）运行上述项目，打开浏览器并在网址栏输入 http://127.0.0.1:8000/book_id/20241001/，其显示内容如图5-20所示。

（6）在浏览器中的网址栏输入 http://127.0.0.1:8000/book_info/，其显示内容如图5-21所示。

图 5-20　浏览器中的显示内容　　　　图 5-21　浏览器中的显示内容

5.4.9　动态构建请求 URL

当视图函数所对应的请求 URL 被大量修改时，如果通过动态构建请求 URL，则无须手动去更改其他地方已经使用该视图函数所对应的请求 URL。

可以通过 django.shortcuts 模块中的 reverse() 函数动态地构建指定视图函数的请求 URL，其语法格式如下：

```
reverse(viewname, kwargs)
```

其中，参数 viewname 表示 URL 命名，参数 kwargs 表示请求 URL 中的参数。

下面通过一个示例，演示一下如何动态地构建请求 URL。

（1）创建名为 djangoProject 的 Django 项目。

（2）打开命令提示符窗口，进入项目所在的根目录中，激活虚拟环境，并输入命令 python manage.py startapp book，创建名为 book 的应用，如图5-22所示。

图 5-22　创建应用

（3）打开应用 book 中的文件 views.py，代码如下：

```
#资源包\Code\chapter5\5.4\4\djangoProject\book\views.py
from django.shortcuts import redirect, reverse
from django.http import HttpResponse
def index(request):
```

```python
    username = request.GET.get('username')
    if username:
        return HttpResponse(f'这里是书籍详情的首页')
    else:
        return redirect(reverse('book_detail', kwargs = {'book_id': '20241001'}))
def book_detail(request, book_id):
    return HttpResponse(f'书籍的id为{book_id}')
```

(4) 打开应用 djangoProject 中的文件 urls.py,代码如下：

```python
#资源包\Code\chapter5\5.4\4\djangoProject\djangoProject\urls.py
from django.urls import path
from book import views
urlpatterns = [
    path('', views.index, name = 'index'),
    path('detail/<book_id>/', views.book_detail, name = 'book_detail')
]
```

(5) 运行上述项目,打开浏览器并在网址栏输入 http://127.0.0.1:8000/?username=oldxia,其显示内容如图 5-23 所示。

(6) 在浏览器中的网址栏输入 http://127.0.0.1:8000/,其显示内容如图 5-24 所示。

图 5-23　浏览器中的显示内容　　　　图 5-24　浏览器中的显示内容

5.4.10　路由分发

在 Django 中,可以通过 django.urls 模块中的 include()函数来实现路由分发,其通常用于将应用的路由组织在一起,或者将特定的路由模式分配给特定的应用。

include()函数具有以下 3 种常用的语法格式：

```
include(pattern, namespace)
```

其中,参数 pattern 表示应用中的模块,参数 namespace 表示实例命名空间。

```
include(pattern, app_namespace, namespace)
```

其中,参数 pattern 表示应用中的模块,参数 app_namespace 表示应用命名空间,参数 namespace 表示实例命名空间。

```
include(pattern_list)
```

其中,参数 pattern_list 表示 URL 映射组成的列表。

这里重点讲解 include()函数中的参数所表示的应用命名空间和实例命名空间。

1. 应用命名空间

由于在多个应用之间可能会存在同名的 URL,所以为了避免动态地构建请求 URL 时产生混淆,可以通过应用命名空间进行区分。定义应用命名空间非常简单,只需在应用中的文件 urls.py 中定义变量 app_name,便可通过该变量指定当前应用的命名空间。

2. 实例命名空间

由于一个应用可以创建多个 URL 映射,所以就会产生一个问题,即在动态构建请求 URL 时,可能产生混淆。为了避免这个问题,可以使用实例命名空间。

除此之外,在使用实例命名空间时,必须指定应用命名空间,否则程序会报错。

下面通过一个示例演示一下如何使用路由分发。

(1) 创建名为 djangoProject 的 Django 项目。

(2) 打开命令提示符窗口,进入项目所在的根目录中,激活虚拟环境,并输入命令 python manage.py startapp front、python manage.py startapp cms、python manage.py startapp movie 和 python manage.py startapp book,创建名为 front、cms、movie 和 book 的应用,如图 5-25 所示。

图 5-25 创建应用

(3) 分别在应用 front、cms 和 movie 中创建文件 urls.py,如图 5-26 所示。

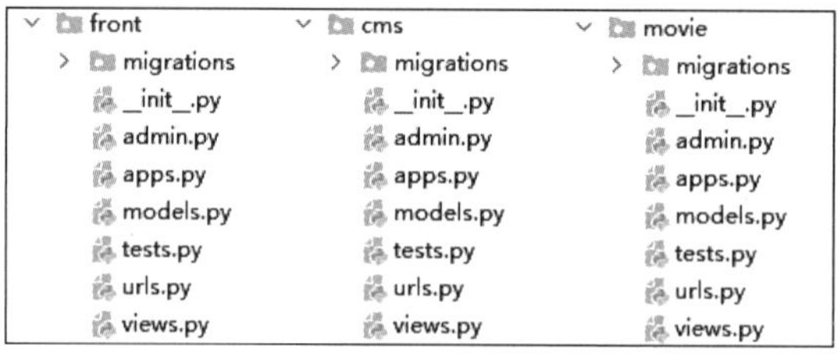

图 5-26 在应用中创建文件 urls.py

(4) 打开应用 front 中的文件 views.py,代码如下:

```
# 资源包\Code\chapter5\5.4\5\djangoProject\front\views.py
from django.http import HttpResponse
from django.shortcuts import redirect, reverse
def index(request):
    username = request.GET.get('username')
    if username:
        return HttpResponse('这里是网站的首页')
    else:
        # 使用应用命名空间
        return redirect(reverse('front:login'))
def login(request):
    return HttpResponse('这里是网站的登录页面')
```

(5) 打开应用 cms 中的文件 views.py,代码如下:

```
# 资源包\Code\chapter5\5.4\5\djangoProject\cms\views.py
from django.http import HttpResponse
```

```python
from django.shortcuts import redirect, reverse
def index(request):
    username = request.GET.get('username')
    if username:
        return HttpResponse('这里是 CMS 的首页')
    else:
        # 获取当前实例命名空间
        current_namespace = request.resolver_match.namespace
        # 使用实例命名空间
        return redirect(reverse(f'{current_namespace}:login'))
def login(request):
    return HttpResponse('这里是 CMS 的登录页面')
```

（6）打开应用 movie 中的文件 views.py，代码如下：

```python
# 资源包\Code\chapter5\5.4\5\djangoProject\movie\views.py
from django.http import HttpResponse
from django.shortcuts import redirect, reverse
def index(request):
    username = request.GET.get('username')
    if username:
        return HttpResponse('这里是电影的首页')
    else:
        # 使用应用命名空间
        return redirect(reverse('movie:login'))
def login(request):
    return HttpResponse('这里是电影的登录页面')
```

（7）打开应用 book 中的文件 views.py，代码如下：

```python
# 资源包\Code\chapter5\5.4\5\djangoProject\book\views.py
from django.http import HttpResponse
def book(request):
    return HttpResponse("这里是书籍的首页")
def book_detail(request):
    return HttpResponse("这里是书籍详情的页面")
```

（8）打开应用 front 中的文件 urls.py，代码如下：

```python
# 资源包\Code\chapter5\5.4\5\djangoProject\front\urls.py
from django.urls import path
from front import views
# 通过变量 app_name 设置应用命名空间
app_name = 'front'
urlpatterns = [
    path('', views.index, name='index'),
    path('login/', views.login, name='login'),
]
```

（9）打开应用 cms 中的文件 urls.py，代码如下：

```python
# 资源包\Code\chapter5\5.4\5\djangoProject\cms\urls.py
from django.urls import path
from cms import views
```

```python
# 通过变量 app_name 设置应用命名空间
app_name = 'cms'
urlpatterns = [
    path('', views.index, name = 'index'),
    path('login/', views.login, name = 'login')
]
```

(10) 打开应用 movie 中的文件 urls.py，代码如下：

```python
# 资源包\Code\chapter5\5.4\5\djangoProject\movie\urls.py
from django.urls import path
from movie import views
urlpatterns = [
    path('', views.index, name = 'index'),
    path('login/', views.login, name = 'login')
]
```

(11) 打开应用 djangoProject 中的文件 urls.py，代码如下：

```python
# 资源包\Code\chapter5\5.4\5\djangoProject\djangoProject\urls.py
from django.urls import path, include
from book import views
urlpatterns = [
    path('', include('front.urls')),
    # 在同一个应用下，具有两个实例，为了避免混淆，可以使用实例命名空间
    path('cms1/', include('cms.urls', namespace = 'cms1')),
    path('cms2/', include('cms.urls', namespace = 'cms2')),
    path('movie/', include(('movie.urls', 'movie'))),
    path('book/', include([
        path('', views.book),
        path('detail/', views.book_detail)
    ]))
]
```

(12) 运行上述项目，打开浏览器并在网址栏输入 http://127.0.0.1:8000/，其显示内容如图 5-27 所示。

(13) 在浏览器中的网址栏输入 http://127.0.0.1:8000/?username=oldxia，其显示内容如图 5-28 所示。

图 5-27　浏览器中的显示内容(1)

图 5-28　浏览器中的显示内容(2)

(14) 在浏览器中的网址栏输入 http://127.0.0.1:8000/cms1/login/，其显示内容如图 5-29 所示。

(15) 在浏览器中的网址栏输入 http://127.0.0.1:8000/cms2/?username=oldxia，其显示内容如图 5-30 所示。

图 5-29　浏览器中的显示内容(3) 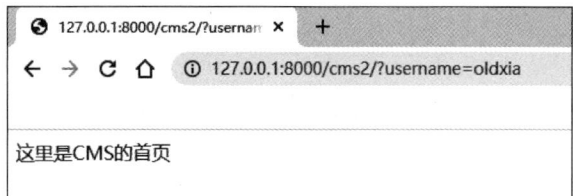　　　　图 5-30　浏览器中的显示内容(4)

（16）在浏览器中的网址栏输入 http://127.0.0.1:8000/movie/login/，其显示内容如图 5-31 所示。

（17）在浏览器中的网址栏输入 http://127.0.0.1:8000/movie/?username=oldxia，其显示内容如图 5-32 所示。

图 5-31　浏览器中的显示内容(5) 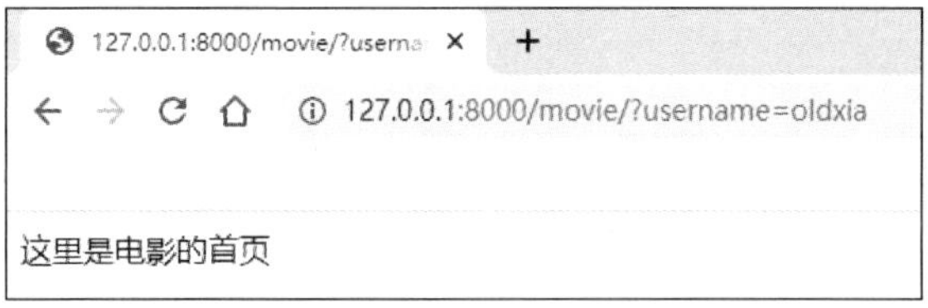　　　　图 5-32　浏览器中的显示内容(6)

（18）在浏览器中的网址栏输入 http://127.0.0.1:8000/book/，其显示内容如图 5-33 所示。

（19）在浏览器中的网址栏输入 http://127.0.0.1:8000/book/detail/，其显示内容如图 5-34 所示。

图 5-33　浏览器中的显示内容(7) 　　　　图 5-34　浏览器中的显示内容(8)

5.4.11　路由转换器

路由转换器主要用于对请求 URL 的参数的类型进行限制，其可以分为内置转换器和自定义转换器。

1. 内置转换器

可以通过将请求 URL 路径中的一部分标记为<converter:variable_name>的方式添加内置转换器，其中，converter 表示限制的规则，包括 str（默认转换器，接受任何不包含斜杠的文本）、int（接受正整数）、slug（接受连字符、下画线、数字或字母）、path（类似 string，但可以接受包含斜杠的文本）和 uuid（接受 uuid 字符串）。

下面通过一个示例演示一下如何使用内置转换器。

（1）创建名为 djangoProject 的 Django 项目。

(2)打开命令提示符窗口,进入项目所在的根目录中,激活虚拟环境,并输入命令 python manage.py startapp book,创建名为 book 的应用,如图 5-35 所示。

```
E:\Python全栈开发\djangoProject>workon django_env
(django_env) E:\Python全栈开发\djangoProject>python manage.py startapp book
```

图 5-35　创建应用

(3)打开应用 book 中的文件 views.py,代码如下:

```python
# 资源包\Code\chapter5\5.4\6\djangoProject\book\views.py
from django.http import HttpResponse
def book_int(request, test_int):
    return HttpResponse(f"书籍的id为{test_int}")
def book_str(request, test_str):
    return HttpResponse(f"书籍的id为{test_str}")
def book_slug(request, test_slug):
    return HttpResponse(f"书籍的id为{test_slug}")
def book_path(request, test_path):
    return HttpResponse(f"书籍的id为{test_path}")
def book_uuid(request, test_uuid):
    return HttpResponse(f"书籍的id为{test_uuid}")
```

(4)打开应用 djangoProject 中的文件 urls.py,代码如下:

```python
# 资源包\Code\chapter5\5.4\6\djangoProject\djangoProject\urls.py
from django.urls import path
from django.http import HttpResponse
from book import views
def index(request):
    return HttpResponse('这是首页')
urlpatterns = [
    path('', index),
    path('book_int/<int:test_int>/', views.book_int),
    path('book_str/<str:test_str>/', views.book_str),
    path('book_slug/<slug:test_slug>/', views.book_slug),
    path('book_path/<path:test_path>/', views.book_path),
    path('book_uuid/<uuid:test_uuid>/', views.book_uuid),
]
```

(5)运行上述项目,打开浏览器并在网址栏输入 http://127.0.0.1:8000/book_int/20241001/,其显示内容如图 5-36 所示。

(6)在浏览器中的网址栏输入 http://127.0.0.1:8000/book_str/oldxia/,其显示内容如图 5-37 所示。

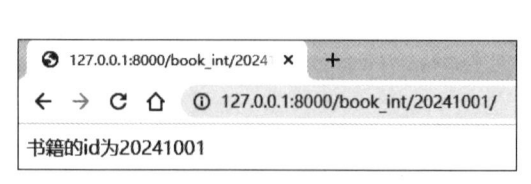

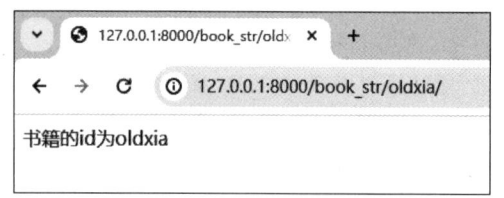

图 5-36　浏览器中的显示内容(1)　　　图 5-37　浏览器中的显示内容(2)

(7)在浏览器中的网址栏输入 http://127.0.0.1:8000/book_slug/oldxia-20241001_

2244999777/,其显示内容如图 5-38 所示。

图 5-38　浏览器中的显示内容(3)

(8) 在浏览器中的网址栏输入 http://127.0.0.1:8000/book_path/oldxia/13309861086/,其显示内容如图 5-39 所示。

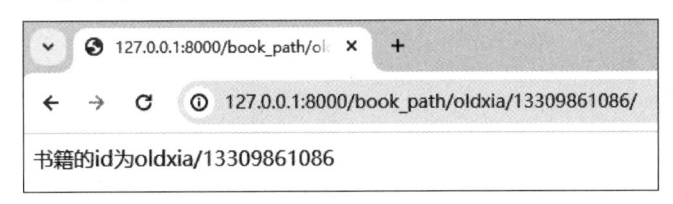

图 5-39　浏览器中的显示内容(4)

(9) 在浏览器中的网址栏输入 http://127.0.0.1:8000/book_uuid/123e4567-e89b-4123-b567-0987654321ab/,其显示内容如图 5-40 所示。

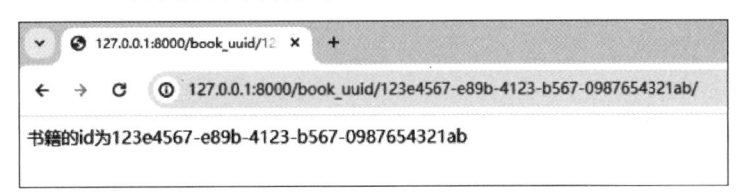

图 5-40　浏览器中的显示内容(5)

2. 自定义转换器

创建自定义转换器分为 3 步。

(1) 创建自定义转换器类,并在该类中实现相关属性和方法,如表 5-4 所示。

表 5-4　自定义转换器类中的属性和方法

属　性	描　述
regex	匹配请求 URL 参数的正则表达式
方　法	描　述
to_python()	当匹配到请求 URL 的参数后,该方法会将其返回值传递到该请求 URL 所对应的视图函数中
to_url()	当其他视图函数使 reverse() 函数时,该方法会对传入的请求 URL 参数进行处理并返回

(2) 在自定义转换器类中,可以使用 django.urls 模块中的 register_converter() 函数将该自定义转化器注册到 Django 中。

(3) 在项目同名的应用中的文件 _init_.py 内导入该自定义转换器类。

下面通过一个示例演示一下如何自定义转换器。

(1) 创建名为 djangoProject 的 Django 项目。

(2) 打开命令提示符窗口,进入项目所在的根目录中,激活虚拟环境,并输入命令

python manage.py startapp book，创建名为 book 的应用，如图 5-41 所示。

图 5-41　创建应用

（3）在应用 book 中创建文件 urls.py 和 converters.py，如图 5-42 所示。

图 5-42　创建文件 urls.py 和 converters.py

（4）打开应用 book 中的文件 converters.py，用于自定义转换器类，代码如下：

```
# 资源包\Code\chapter5\5.4\7\djangoProject\book\converters.py
from django.urls import register_converter
# 自定义转换器类
class CategoryConverter(object):
    regex = r'\w+|(\w+\+\w+)+'
    def to_python(self, value):
        result = value.split('+')
        return result
    def to_url(self, value):
        if isinstance(value, list):
            result = '+'.join(value)
            return result
        else:
            raise RuntimeError('参数必须为【list】类型')
# 将自定义转换器类注册到 Django 中
register_converter(CategoryConverter, 'cate')
```

（5）打开应用 djangoProject 中的文件 __init__.py，代码如下：

```
# 资源包\Code\chapter5\5.4\7\djangoProject\djangoProject\__init__.py
from book import converters
```

（6）打开应用 book 中的文件 views.py，代码如下：

```
# 资源包\Code\chapter5\5.4\7\djangoProject\book\views.py
from django.http import HttpResponse
from django.shortcuts import reverse
def book_cate(request, categories):
    print(reverse('list', kwargs = {'categories': categories}))
    return HttpResponse(f"该书籍的分类为{categories}")
```

（7）打开应用 book 中的文件 urls.py，代码如下：

```
# 资源包\Code\chapter5\5.4\7\djangoProject\book\urls.py
from django.urls import path
from book import views
```

```
urlpatterns = [
    path('cate/<cate:categories>/', views.book_cate, name='list'),
]
```

(8) 打开应用 djangoProject 中的文件 urls.py，代码如下：

```
# 资源包\Code\chapter5\5.4\7\djangoProject\djangoProject\urls.py
from django.urls import path, include
urlpatterns = [
    path('book/', include('book.urls'))
]
```

(9) 运行上述程序，打开浏览器并在网址栏输入 http://127.0.0.1:8000/book/cate/python+django+flask/，其显示内容如图 5-43 所示。

图 5-43　浏览器中的显示内容

此时，PyCharm 中的显示结果如图 5-44 所示。

图 5-44　PyCharm 中的输出结果

5.4.12　限制请求方法

在 Django 中，可以通过 django.views.decorators.http 模块中的内置装饰器限制视图函数的请求方法，常用的装饰器包括 require_GET、require_POST 和 require_http_methods。

下面通过一个示例演示一下如何限制视图函数的请求方法。

(1) 创建名为 djangoProject 的 Django 项目。

(2) 打开命令提示符窗口，进入项目所在的根目录中，激活虚拟环境，并输入命令 python manage.py startapp book，创建名为 book 的应用，如图 5-45 所示。

图 5-45　创建应用

(3) 在项目根目录下的文件夹 templates 中创建模板文件 index.html 文件(模板技术将在后续章节为读者详细讲解),代码如下:

```html
# 资源包\Code\chapter5\5.4\8\djangoProject\templates\index.html
<html lang="en">
    <head>
        <meta charset="UTF-8">
        <title>Title</title>
    </head>
    <body>
        <form action="{% url 'post' %}" method="post">
            <!-- CSRF 防御的相关知识点后续章节将为读者详细讲解 -->
            {% csrf_token %}
            <input type="submit" value="模拟 POST 提交">
        </form>
    </body>
</html>
```

(4) 打开应用 book 中的文件 views.py,代码如下:

```python
# 资源包\Code\chapter5\5.4\8\djangoProject\book\views.py
from django.http import HttpResponse
from django.shortcuts import render
from django.views.decorators.http import require_GET, require_http_methods
def index(request):
    return render(request, 'index.html')
# @require_GET 等价于@require_http_methods(['GET'])
@require_GET
def post_test(request):
    return HttpResponse("通过 POST 方法访问!")
```

(5) 打开应用 djangoProject 中的文件 urls.py,代码如下:

```python
# 资源包\Code\chapter5\5.4\8\djangoProject\djangoProject\urls.py
from django.urls import path
from book import views
urlpatterns = [
    path('', views.index),
    path('post_test/', views.post_test, name='post'),
]
```

(6) 运行上述项目,打开浏览器并在网址栏输入 http://127.0.0.1:8000/,其显示内容如图 5-46 所示。

图 5-46 浏览器中的显示内容

(7) 单击"模拟 POST 提交"按钮,其显示内容如图 5-47 所示,即 HTTP 状态码 405,表示请求中指定的方法不被允许。

图 5-47　浏览器中的显示内容

5.5 模板

在 Django 中,模板是用于帮助开发者快速生成前端页面的工具,可以大幅降低代码的复杂度和维护成本。

Django 中内置了一款模板引擎,即 DTL(Django Template Language)。除此之外,Django 还支持第三方模板引擎,如 Jinja2 等,但由于 DTL 可以和 Django 达到无缝衔接而不会产生不兼容的情况,因此建议广大读者优先学习该模板引擎。

5.5.1 渲染模板

在 Django 中提供了两种常用的渲染方式,即 render_to_string()函数和 render()函数。

1. render_to_string()函数

通过 django.template.loader 模块中的 render_to_string()函数进行模板的渲染,其语法格式如下:

```
render_to_string(template_name, context)
```

其中,参数 template_name 表示待渲染模板的名称,参数 context 表示模板中的变量,并且必须为字典类型。

2. render()函数

通过 django.shortcuts 模块中的 render()函数进行模板的渲染,其语法格式如下:

```
render(request, template_name, context)
```

其中,参数 request 表示 HttpRequest 对象,参数 template_name 表示待渲染模板的名称,参数 context 表示模板中的变量,并且必须为字典类型。

下面通过一个示例演示一下如何渲染模板。

(1)创建名为 djangoProject 的 Django 项目。

(2)在项目根目录下的文件夹 templates 中创建模板文件 render_to_string.html,代码如下:

```
# 资源包\Code\chapter5\5.5\1\djangoProject\templates\render_to_string.html
<!DOCTYPE html>
```

```html
<html lang="en">
    <head>
        <meta charset="UTF-8">
        <title>Title</title>
        <style>
            h1 {
                color: red;
            }
        </style>
    </head>
    <body>
        <h1>这是使用render_to_string()函数渲染的页面</h1>
    </body>
</html>
```

(3) 在项目根目录下的文件夹 templates 中创建模板文件 myrender.html，代码如下：

```html
#资源包\Code\chapter5\5.5\1\djangoProject\templates\myrender.html
<!DOCTYPE html>
<html lang="en">
    <head>
        <meta charset="UTF-8">
        <title>Title</title>
        <style>
            h1 {
                color: green;
            }
        </style>
    </head>
    <body>
        <h1>这是使用render()函数渲染的页面</h1>
    </body>
</html>
```

(4) 打开应用 djangoProject 中的文件 urls.py，代码如下：

```python
#资源包\Code\chapter5\5.5\1\djangoProject\DjangoProject\urls.py
from django.contrib import admin
from django.urls import path
from django.template.loader import render_to_string
from django.shortcuts import render
from django.http import HttpResponse
def renderToString(request):
    html = render_to_string('render_to_string.html')
    return HttpResponse(html)
def myRender(request):
    return render(request, 'myrender.html')
urlpatterns = [
    path('admin/', admin.site.urls),
    path('rendertostring/', renderToString),
    path('myrender/', myRender),
]
```

(5) 运行上述项目，打开浏览器并在网址栏输入 http://127.0.0.1:8000/rendertostring/，其显示内容如图 5-48 所示。

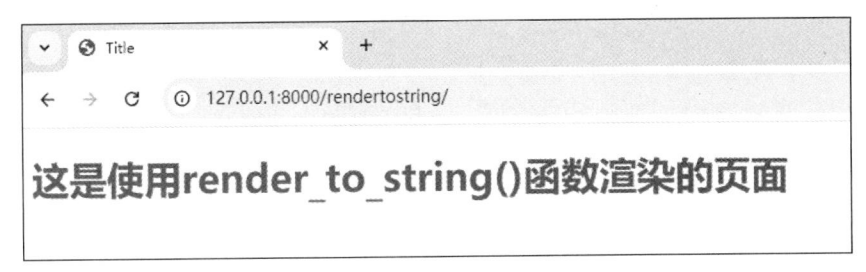

图 5-48　浏览器中的显示内容(1)

（6）在浏览器中的网址栏输入 http://127.0.0.1:8000/myrender/，其显示内容如图 5-49 所示。

图 5-49　浏览器中的显示内容(2)

5.5.2　模板位置

在 Django 中，模板位置通过项目同名的应用中的配置文件 settings.py 的 TEMPLATES 配置中的 DIRS 进行设置，在默认情况下，模板文件会存放在项目根目录下的文件夹 templates 之中。

此外，通过将 TEMPLATES 配置中的 APP_DIRS 的值设置为 True，并将指定的应用添加到配置文件中的变量 INSTALLED_APPS 之中，即可完成应用注册操作，然后在指定的应用中创建文件夹 templates 并添加相应的模板文件，即可使模板引擎进入每个已注册的应用中查找模板。

模板查找的顺序，一是当 DIRS 中有模板路径，模板引擎会优先查找当前路径下的模板；二是在 APP_DIRS 的值为 True 时，模板引擎会开始查找已注册应用的模板。

下面通过一个示例，演示一下如何配置模板位置。

（1）创建名为 djangoProject 的 Django 项目。

（2）打开命令提示符窗口，进入项目所在的根目录中，激活虚拟环境，并输入命令 python manage.py startapp front 和 python manage.py startapp cms，创建名为 front 和 cms 的应用，如图 5-50 所示。

图 5-50　创建应用

（3）分别在应用 front 和 cms 中创建文件夹 templates 和文件 urls.py，如图 5-51 所示。

（4）在项目根目录下的文件夹 templates 中创建模板文件 index.html，代码如下：

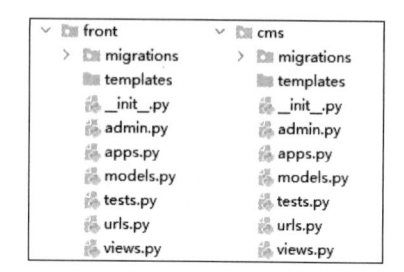

图 5-51　在应用中创建文件夹 templates 和文件 urls.py

```
# 资源包\Code\chapter5\5.5\2\djangoProject\templates\index.html
<!DOCTYPE html>
<html lang="en">
    <head>
        <meta charset="UTF-8">
        <title>Title</title>
    </head>
    <body>
        <h1>这是网站的首页</h1>
    </body>
</html>
```

（5）在应用 front 中的文件夹 templates 中创建模板文件 front_login.html，代码如下：

```
# 资源包\Code\chapter5\5.5\2\djangoProject\front\templates\front_login.html
<!DOCTYPE html>
<html lang="en">
    <head>
        <meta charset="UTF-8">
        <title>Title</title>
    </head>
    <body>
        <h1>这是 Front 的登录页面</h1>
    </body>
</html>
```

（6）在应用 cms 中的文件夹 templates 中创建模板文件 cms_login.html，代码如下：

```
# 资源包\Code\chapter5\5.5\2\djangoProject\cms\templates\cms_login.html
<!DOCTYPE html>
<html lang="en">
    <head>
        <meta charset="UTF-8">
        <title>Title</title>
    </head>
    <body>
        <h1>这是 CMS 的登录页面</h1>
    </body>
</html>
```

（7）打开应用 djangoProject 中的配置文件 settings.py，将应用 front 和 cms 添加到变量 INSTALLED_APPS 中，完成应用注册，代码如下：

```python
# 资源包\Code\chapter5\5.5\2\djangoProject\djangoProject\settings.py
INSTALLED_APPS = [
    'django.contrib.admin',
    'django.contrib.auth',
    'django.contrib.contenttypes',
    'django.contrib.sessions',
    'django.contrib.messages',
    'django.contrib.staticfiles',
    'front',
    'cms',
]
```

（8）打开应用 front 中的文件 views.py，代码如下：

```python
# 资源包\Code\chapter5\5.5\2\djangoProject\front\views.py
from django.shortcuts import render
def login(request):
    return render(request, 'front_login.html')
```

（9）打开应用 cms 中的文件 views.py，代码如下：

```python
# 资源包\Code\chapter5\5.5\2\djangoProject\cms\views.py
from django.shortcuts import render
def login(request):
    return render(request, 'cms_login.html')
```

（10）打开应用 front 中的文件 urls.py，代码如下：

```python
# 资源包\Code\chapter5\5.5\2\djangoProject\front\urls.py
from django.urls import path
from front import views
urlpatterns = [
    path('', views.login),
]
```

（11）打开应用 cms 中的文件 urls.py，代码如下：

```python
# 资源包\Code\chapter5\5.5\2\djangoProject\cms\urls.py
from django.urls import path
from cms import views
urlpatterns = [
    path('login/', views.login),
]
```

（12）打开应用 djangoProject 中的文件 urls.py，代码如下：

```python
# 资源包\Code\chapter5\5.5\2\djangoProject\djangoProject\urls.py
from django.urls import path, include
from django.shortcuts import render
def index(request):
    return render(request, 'index.html')
urlpatterns = [
    path('', index),
    path('login/', include('front.urls')),
    path('cms/', include('cms.urls'))
]
```

(13) 运行上述项目，打开浏览器并在网址栏输入 http://127.0.0.1:8000/，其显示内容如图 5-52 所示。

(14) 在浏览器中的网址栏输入 http://127.0.0.1:8000/login/，其显示内容如图 5-53 所示。

图 5-52　浏览器中的显示内容(1)　　　　图 5-53　浏览器中的显示内容(2)

(15) 在浏览器中的网址栏输入 http://127.0.0.1:8000/cms/login/，其显示内容如图 5-54 所示。

图 5-54　浏览器中的显示内容(3)

5.5.3　模板变量

在模板中可以通过标签{{ variable_name }}来获取 render()函数所传递的参数，并在渲染模板时将其解析成对应的值。

下面通过一个示例演示一下如何使用模板变量。

(1) 创建名为 djangoProject 的 Django 项目。

(2) 打开应用 djangoProject 中的文件 urls.py，代码如下：

```python
# 资源包\Code\chapter5\5.5\3\djangoProject\djangoProject\urls.py
from django.urls import path
from django.shortcuts import render
class Course():
    def __init__(self, name, teacher):
        self.name = name
        self.teacher = teacher
def index(request):
    course = Course('Django', 'oldxia')
    context = {
        # 字符串
        'username': 'oldxia',
        # 字典
        'person': {
            'username': 'oldxia',
            'age': 38,
        },
```

```
        # 类
        'course': course,
        # 列表
        'language': ['Python', 'Java', 'PHP'],
    }
    return render(request, 'index.html', context = context)
urlpatterns = [
    path('', index),
]
```

(3) 在项目根目录下的文件夹 templates 中创建模板文件 index.html,代码如下:

```
# 资源包\Code\chapter5\5.5\3\djangoProject\templates\index.html
<!DOCTYPE html>
<html lang = "en">
    <head>
        <meta charset = "UTF-8">
        <title>Title</title>
    </head>
    <body>
        <h1>{{ username }}</h1>
        <h1>{{ person.age }}</h1>
        <h1>{{ course.name }}</h1>
        <h1>{{ language.0 }}</h1>
    </body>
</html>
```

(4) 运行上述项目,打开浏览器并在网址栏输入 http://127.0.0.1:8000/,其显示内容如图 5-55 所示。

图 5-55　浏览器中的显示内容

5.5.4　模板中的控制结构

Django 提供了多种控制结构,用于改变模板的渲染流程。

1. 选择结构

1) 单分支选择结构

通过标签{% if %}…{% endif %}来实现单分支选择结构。

2）双分支选择结构

通过标签{% if %}…{% else %}…{% endif %}来实现双分支选择结构。

3）多分支选择结构

通过标签{% if %}…{% elif %}…{% else %}…{% endif %}来实现多分支选择结构。

4）选择结构嵌套

对上述3种选择结构进行相互嵌套，即可用于表达更加复杂的选择结构。

下面通过一个示例演示一下如何使用选择结构。

(1) 创建名为 djangoProject 的 Django 项目。

(2) 打开应用 djangoProject 中的文件 urls.py，代码如下：

```python
# 资源包\Code\chapter5\5.5\4\djangoProject\djangoProject\urls.py
from django.urls import path
from django.shortcuts import render
# 单分支选择结构
def single_select(request):
    context = {
        'username': 'oldxia',
    }
    return render(request, 'single_select.html', context = context)
# 双分支选择结构
def dual_select(request):
    context = {
        'username': 'xzd',
    }
    return render(request, 'dual_select.html', context = context)
# 多分支选择结构
def multi_select(request):
    context = {
        'score': 95,
    }
    return render(request, 'multi_select.html', context = context)
# 选择结构嵌套
def nested_select(request):
    context = {
        'score': 36,
    }
    return render(request, 'nested_select.html', context = context)
urlpatterns = [
    path('single_select/', single_select),
    path('dual_select/', dual_select),
    path('multi_select/', multi_select),
    path('nested_select/', nested_select),
]
```

(3) 在项目根目录下的文件夹 templates 中创建模板文件 single_select.html，代码如下：

```html
# 资源包\Code\chapter5\5.5\4\djangoProject\templates\single_select.html
<!DOCTYPE html>
< html lang = "en">
```

```html
    <head>
        <meta charset="UTF-8">
        <title>Title</title>
    </head>
    <body>
        {% if username == 'oldxia' %}
            <h1>欢迎用户{{ username }}登录网站</h1>
        {% endif %}
    </body>
</html>
```

(4) 在项目根目录下的文件夹 templates 中创建模板文件 dual_select.html,代码如下：

```html
#资源包\Code\chapter5\5.5\4\djangoProject\templates\dual_select.html
<!DOCTYPE html>
<html lang="en">
    <head>
        <meta charset="UTF-8">
        <title>Title</title>
    </head>
    <body>
        {% if username == 'oldxia' %}
            <h1>欢迎用户{{ username }}登录网站</h1>
        {% else %}
            <h1>用户{{ username }}不是本网站用户,无法登录!</h1>
        {% endif %}
    </body>
</html>
```

(5) 在项目根目录下的文件夹 templates 中创建模板文件 multi_select.html,代码如下：

```html
#资源包\Code\chapter5\5.5\4\djangoProject\templates\multi_select.html
<!DOCTYPE html>
<html lang="en">
    <head>
        <meta charset="UTF-8">
        <title>Title</title>
    </head>
    <body>
        {% if score >= 90 %}
            <h1>您的分数{{ score }}分,成绩优秀</h1>
        {% elif score >= 70 %}
            <h1>您的分数{{ score }}分,成绩良好</h1>
        {% elif score >= 60 %}
            <h1>您的分数{{ score }}分,成绩及格</h1>
        {% else %}
            <h1>您的分数{{ score }}分,成绩不及格</h1>
        {% endif %}
    </body>
</html>
```

(6) 在项目根目录下的文件夹 templates 中创建模板文件 nested_select.html,代码如下：

```html
# 资源包\Code\chapter5\5.5\4\djangoProject\templates\nested_select.html
<!DOCTYPE html>
<html lang="en">
    <head>
        <meta charset="UTF-8">
        <title>Title</title>
    </head>
    <body>
        {% if score >= 60 %}
            <h1>您的分数{{ score }}分,考试及格</h1>
        {% else %}
            {% if score >= 45 %}
                <h1>您的分数{{ score }}分,考试不及格,但可以参加补考</h1>
            {% else %}
                <h1>您的分数{{ score }}分,考试不及格,并且不可以参加补考</h1>
            {% endif %}
        {% endif %}
    </body>
</html>
```

（7）运行上述项目,打开浏览器并在网址栏输入 http://127.0.0.1:8000/single_select/,其显示内容如图 5-56 所示。

图 5-56　浏览器中的显示内容(1)

（8）在浏览器中的网址栏输入 http://127.0.0.1:8000/dual_select/,其显示内容如图 5-57 所示。

图 5-57　浏览器中的显示内容(2)

（9）在浏览器中的网址栏输入 http://127.0.0.1:8000/multi_select/,其显示内容如图 5-58 所示。

（10）在浏览器中的网址栏输入 http://127.0.0.1:8000/nested_select/,其显示内容如图 5-59 所示。

2. 循环结构

在模板中,可以通过标签{% for %}…{% endfor %}、标签{% for %}…{% else %}…{% endfor %}或{% for %}…{% empty %}…{% endfor %}来实现循环结构。

图 5-58　浏览器中的显示内容(3)

图 5-59　浏览器中的显示内容(4)

此外,在循环结构中还包含多个内置的循环变量,主要用于获取当前遍历的状态,具体如表 5-5 所示。

表 5-5　内置循环变量

内置循环变量	描述
forloop.counter	当前迭代的索引,开始的索引从 1 开始
forloop.counter0	当前迭代的索引,开始的索引从 0 开始
forloop.revcounter	当前迭代的反向索引,最后一个索引从 1 开始
forloop.revcounter0	当前迭代的反向索引,最后一个索引从 0 开始
forloop.first	是否是第 1 次迭代,返回值为 True 或 False
loop.last	是否是最后一次迭代,返回值为 True 或 False
loop.length	序列的长度

下面通过一个示例演示一下如何使用循环结构。

(1) 创建名为 djangoProject 的 Django 项目。

(2) 打开应用 djangoProject 中的文件 urls.py,代码如下:

```
# 资源包\Code\chapter5\5.5\5\djangoProject\djangoProject\urls.py
from django.urls import path
from django.shortcuts import render
def index(request):
    context = {
        'books': [
            {
                "name": "《Python 全栈开发——基础入门》",
                "author": "夏正东",
                "price": 79
            },
            {
                "name": "《Python 全栈开发——高阶编程》",
                "author": "夏正东",
                "price": 89
```

```
        },
        {
            "name": "《Python 全栈开发——数据分析》",
            "author": "夏正东",
            "price": 79
        },
        {
            "name": "《Python 全栈开发——Web 编程》",
            "author": "夏正东",
            "price": "待定"
        }
    ]
}
    return render(request, 'index.html', context = context)
urlpatterns = [
    path('', index)
]
```

(3) 在项目根目录下的文件夹 templates 中创建模板文件 index.html，代码如下：

```
#资源包\Code\chapter5\5.5\5\djangoProject\templates\index.html
<!DOCTYPE html>
<html lang = "en">
    <head>
        <meta charset = "UTF-8">
        <title>Title</title>
    </head>
    <body>
        <table border = "1" cellspacing = "0">
            <tr>
                <th>序号 0</th>
                <th>序号 1</th>
                <th>书名</th>
                <th>作者</th>
                <th>价格</th>
            </tr>
            {% for book in books %}
                {% if forloop.first %}
                    <tr style = "background: yellow">
                {% elif forloop.last %}
                    <tr style = "background: greenyellow">
                {% else %}
                    <tr>
                {% endif %}
                    <td>{{ forloop.counter0 }}</td>
                    <td>{{ forloop.counter }}</td>
                    <td>{{ book.name }}</td>
                    <td>{{ book.author }}</td>
                    <td>{{ book.price }}</td>
                </tr>
            {% endfor %}
        </table>
        <ul>
            {% for comment in comments %}
                <li>comment</li>
```

```
                {% empty %}
                    <li>没有任何评论!</li>
                {% endfor %}
            </ul>
        </body>
</html>
```

（4）运行上述项目，打开浏览器并在网址栏输入 http://127.0.0.1:8000/，其显示内容如图 5-60 所示。

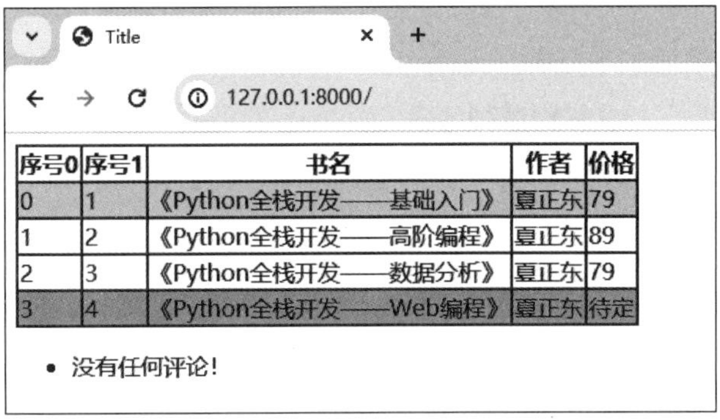

图 5-60　浏览器中的显示内容

5.5.5　模板注释

在模板中，可以通过标签{# … #}添加注释，需要注意的是模板注释不会出现在 HTML 文档之中。

下面通过一个示例演示一下如何使用模板注释。

（1）创建名为 djangoProject 的 Django 项目。

（2）打开应用 djangoProject 中的文件 urls.py，代码如下：

```
#资源包\Code\chapter5\5.5\6\djangoProject\djangoProject\urls.py
from django.urls import path
from django.shortcuts import render
def index(request):
    return render(request, 'index.html')
urlpatterns = [
    path('', index)
]
```

（3）在项目根目录下的文件夹 templates 中创建模板文件 index.html，代码如下：

```
#资源包\Code\chapter5\5.5\6\djangoProject\templates\index.html
<!DOCTYPE html>
<html lang="en">
    <head>
        <meta charset="UTF-8">
        <title>Title</title>
    </head>
    <body>
```

```html
        {# 这是模板注释 #}
        <h1>模板注释</h1>
    </body>
</html>
```

（4）运行上述项目，打开浏览器并在网址栏输入 http://127.0.0.1:8000/，其显示内容如图 5-61 所示。

图 5-61　浏览器中的显示内容

5.5.6　常用标签

1. with 标签

标签 {% with… %}…{% endwith %} 用于定义作用域在该标签内的模板变量。

下面通过一个示例演示一下如何使用 with 标签。

（1）创建名为 djangoProject 的 Django 项目。

（2）打开应用 djangoProject 中的文件 urls.py，代码如下：

```python
# 资源包\Code\chapter5\5.5\7\djangoProject\djangoProject\urls.py
from django.urls import path
from django.shortcuts import render
def index(request):
    return render(request, 'index.html')
urlpatterns = [
    path('', index)
]
```

（3）在项目根目录下的文件夹 templates 中创建模板文件 index.html，代码如下：

```html
# 资源包\Code\chapter5\5.5\7\djangoProject\templates\index.html
<!DOCTYPE html>
<html lang="en">
    <head>
        <meta charset="UTF-8">
        <title>Title</title>
    </head>
    <body>
        {% with website = "http://www.oldxia.com" %}
            <p>老夏学院的网址:{{ website }}</p>
        {% endwith %}
    </body>
</html>
```

（4）运行上述项目，打开浏览器并在网址栏输入 http://127.0.0.1:8000/，其显示内容如图 5-62 所示。

图 5-62 浏览器中的显示内容

2. url 标签

标签{% url…%}用于动态构建请求 URL。

下面通过一个示例演示一下如何使用 url 标签。

(1) 创建名为 djangoProject 的 Django 项目。

(2) 打开应用 djangoProject 中的文件 urls.py,代码如下:

```python
# 资源包\Code\chapter5\5.5\8\djangoProject\djangoProject\urls.py
from django.urls import path
from django.shortcuts import render
def index(request):
    return render(request, 'index.html')
def login(request, username, course):
    if username:
        context = {
            'username': username,
            'course': course,
        }
        return render(request, 'login.html', context=context)
urlpatterns = [
    path('', index),
    path('login/<username>/<course>', login, name='login')
]
```

(3) 在项目根目录下的文件夹 templates 中创建模板文件 index.html,代码如下:

```html
# 资源包\Code\chapter5\5.5\8\djangoProject\templates\index.html
<!DOCTYPE html>
<html lang="en">
    <head>
        <meta charset="UTF-8">
        <title>Title</title>
    </head>
    <body>
        <h1><a href="{% url 'login' username='oldxia' course='Django' %}">单击此处进行登录</a></h1>
    </body>
</html>
```

(4) 在项目根目录下的文件夹 templates 中创建模板文件 login.html,代码如下:

```html
# 资源包\Code\chapter5\5.5\8\djangoProject\templates\login.html
<!DOCTYPE html>
<html lang="en">
    <head>
        <meta charset="UTF-8">
```

```html
        <title>Title</title>
    </head>
    <body>
        <h1>欢迎{{ username }}登录本网站,您学习的课程内容是{{ course }}</h1>
    </body>
</html>
```

(5) 运行上述项目,打开浏览器并在网址栏输入 http://127.0.0.1:8000/,其显示内容如图 5-63 所示。

图 5-63　浏览器中的显示内容(1)

(6) 单击"单击此处进行登录"按钮,页面将跳转至动态构建的请求 URL,其显示内容如图 5-64 所示。

图 5-64　浏览器中的显示内容(2)

3. spaceless 标签

标签{% spaceless %}…{% endspaceless %}用于移除 HTML 中标签与标签之间的空白符。

下面通过一个示例演示一下如何使用 spaceless 标签。

(1) 创建名为 djangoProject 的 Django 项目。

(2) 打开应用 djangoProject 中的文件 urls.py,代码如下:

```python
#资源包\Code\chapter5\5.5\9\djangoProject\djangoProject\urls.py
from django.urls import path
from django.shortcuts import render
def index(request):
    return render(request, 'index.html')
urlpatterns = [
    path('', index)
]
```

(3) 在项目根目录下的文件夹 templates 中创建模板文件 index.html,代码如下:

```html
#资源包\Code\chapter5\5.5\9\djangoProject\templates\index.html
<!DOCTYPE html>
```

```
<html lang = "en">
    <head>
        <meta charset = "UTF-8">
        <title> Title </title>
    </head>
    <body>
        {% spaceless %}
            <p>
                <a href = ""> http://www.oldxia.com </a>
            </p>
        {% endspaceless %}
    </body>
</html>
```

（4）运行上述项目，打开浏览器并在网址栏输入 http://127.0.0.1:8000/，然后在当前页面右击，选择"查看网页源代码"，可以看到 p 标签和 a 标签之间的空白符已经被去除，其显示内容如图 5-65 所示。

图 5-65　浏览器中的显示内容

4. autoescape 标签

标签{% autoescape on[off] %}…{% endautoescape %}用于开启或关闭 HTML 标签的自动转义功能。

下面通过一个示例演示一下如何使用 autoescape 标签。

（1）创建名为 djangoProject 的 Django 项目。

（2）打开应用 djangoProject 中的文件 urls.py，代码如下：

```
# 资源包\Code\chapter5\5.5\10\djangoProject\djangoProject\urls.py
from django.urls import path
from django.shortcuts import render
def index(request):
    context = {
        'website': "< a href = 'http://www.oldxia.com'>老夏学院</a>"
    }
    return render(request, 'index.html', context = context)
urlpatterns = [
    path('', index)
]
```

（3）在项目根目录中的文件夹 templates 中创建模板文件 index.html，代码如下：

```
# 资源包\Code\chapter5\5.5\10\djangoProject\templates\index.html
<!DOCTYPE html>
```

```html
<html lang="en">
    <head>
        <meta charset="UTF-8">
        <title>Title</title>
    </head>
    <body>
        {% autoescape off %}
            <h1>{{ website }}</h1>
        {% endautoescape %}
    </body>
</html>
```

图 5-66 浏览器中的显示内容

（4）运行上述项目，打开浏览器并在网址栏输入 http://127.0.0.1:8000/，其显示内容如图 5-66 所示。

5. verbatim 标签

标签{% verbatim %}…{% endverbatim %}用于关闭 DTL 模板引擎的解析功能。

下面通过一个示例演示一下如何使用 verbatim 标签。

（1）创建名为 djangoProject 的 Django 项目。

（2）打开应用 djangoProject 中的文件 urls.py，代码如下：

```python
# 资源包\Code\chapter5\5.5\11\djangoProject\djangoProject\urls.py
from django.urls import path
from django.shortcuts import render
def index(request):
    context = {
        'website': "<a href='http://www.oldxia.com'>老夏学院</a>"
    }
    return render(request, 'index.html', context=context)
urlpatterns = [
    path('', index)
]
```

（3）在项目根目录下的文件夹 templates 中创建模板文件 index.html，代码如下：

```html
# 资源包\Code\chapter5\5.5\11\djangoProject\templates\index.html
<!DOCTYPE html>
<html lang="en">
    <head>
        <meta charset="UTF-8">
        <title>Title</title>
    </head>
    <body>
        {% verbatim %}
            <h1>{{ website }}</h1>
        {% endverbatim %}
    </body>
</html>
```

（4）运行上述项目，打开浏览器并在网址栏输入 http://127.0.0.1:8000/，其显示内容

6. include 标签

标签{% include… %}用于将一个模板引入另外一个模板中的指定位置。

下面通过一个示例演示一下如何使用 include 标签。

图 5-67　浏览器中的显示内容

（1）创建名为 djangoProject 的 Django 项目。

（2）打开应用 djangoProject 中的文件 urls.py，代码如下：

```python
# 资源包\Code\chapter5\5.5\12\djangoProject\djangoProject\urls.py
from django.urls import path
from django.shortcuts import render
def index(request):
    return render(request, 'index.html')
urlpatterns = [
    path('', index)
]
```

（3）在项目根目录下的文件夹 templates 中创建模板文件 index.html，代码如下：

```html
# 资源包\Code\chapter5\5.5\12\djangoProject\templates\index.html
<!DOCTYPE html>
<html lang="en">
    <head>
        <meta charset="UTF-8">
        <title>Title</title>
    </head>
    <body>
        {% include "header.html" %}
        <form action="">
            名称：<input type="text" name="user"><br><br>
            密码：<input type="password" name="password"><br><br>
            <input type="submit" name="submit" value="提交">
            <input type="reset" name="reset" value="重置">
        </form>
        {% include "footer.html" %}
    </body>
</html>
```

（4）在项目根目录下的文件夹 templates 中创建模板文件 header.html，代码如下：

```html
# 资源包\Code\chapter5\5.5\12\djangoProject\templates\header.html
<h1 style="background: red">这是 header</h1>
```

（5）在项目根目录下的文件夹 templates 中创建模板文件 footer.html，代码如下：

```html
# 资源包\Code\chapter5\5.5\12\djangoProject\templates\footer.html
<h1 style="background: green">这是 footer</h1>
```

（6）运行上述项目，打开浏览器并在网址栏输入 http://127.0.0.1:8000/，其显示内容如图 5-68 所示。

图 5-68　浏览器中的显示内容

5.5.7　模板中的过滤器

在模板中,过滤器相当于一个特殊的函数,主要用于修改和过滤变量的值。过滤器通过管道符号"|"实现。

1. 内置过滤器

在 DTL 中,内置了许多过滤器,如表 5-6 所示。

表 5-6　常用的内置过滤器

过滤器	描述
add	将传递的参数与原值相加
cut	将原值中所有指定的字符串移除
date	将日期按照指定的格式格式化为字符串
default	为原值设置默认值
default_if_none	当原值为 None 时,为其设置默认值
first	返回序列中的第 1 个元素
last	返回序列中的最后一个元素
floatformat	对浮点数进行格式化
join	对原值与传递的参数进行拼接
length	返回序列或字典的长度
lower	将原值中的字符转换为小写
upper	将原值中的字符转换为大写
random	随机返回序列中的值
safe	关闭字符串的自动转义
slice	对原值进行切片操作
striptags	去除原值中的 HTML 标签
truncatechars	按照指定长度对原值进行截断,并在末尾拼接 3 个点
truncatechars_html	按照指定长度对原值中的非 HTML 标签部分进行截断,并在末尾拼接 3 个点

下面通过一个示例演示一下如何使用内置过滤器。

(1) 创建名为 djangoProject 的 Django 项目。

(2) 打开应用 djangoProject 中的文件 urls.py，代码如下：

```python
# 资源包\Code\chapter5\5.5\13\djangoProject\djangoProject\urls.py
from django.urls import path
from django.shortcuts import render
from datetime import datetime
def index(request):
    context = {
        'add_filter': 'django',
        'cut_filter': 'd-j-a-n-g-o',
        'date_filter': datetime.now(),
        'default_filter': '',
        'default_if_none_filter': None,
        'first_filter': ['Python', 'Java', 'Go'],
        'last_filter': ['Python', 'Java', 'Go'],
        'floatformat_filter': 100.656,
        'join_filter': ['Django', 'Flask'],
        'length_filter': ['Django', 'Flask'],
        'lower_filter': 'Hello Django',
        'upper_filter': 'hello django',
        'random_filter': [1, 2, 3, 4, 5, 6, 7],
        'safe_filter': "<a href='http://www.oldxia.com'>老夏学院</a>",
        'slice_filter': 'www.oldxia.com',
        'striptags_filter': '<h1>老夏学院</h1>',
        'truncatechars_filter': '<h1>Python-Django</h1>',
        'truncatechars_htmls_filter': '<h1>Python-Django</h1>',
    }
    return render(request, 'index.html', context=context)
urlpatterns = [
    path('', index),
]
```

(3) 在项目根目录下的文件夹 templates 中创建模板文件 index.html，代码如下：

```html
# 资源包\Code\chapter5\5.5\13\djangoProject\templates\index.html
<!DOCTYPE html>
<html lang="en">
    <head>
        <meta charset="UTF-8">
        <title>Title</title>
    </head>
    <body>
        <!-- 注意：过滤器后的冒号前后不能有空格！ -->
        <p>add: {{add_filter|add:'+Python'}}</p>
        <p>cut: {{cut_filter|cut:'-'}}</p>
        <p>date: {{date_filter|date:'Y-m-d h:i:s'}}</p>
        <p>default: {{default_filter|default:'原值为空值,使用默认值'}}</p>
        <p>default_if_none: {{default_if_none_filter|default:'原值为None,使用默认值'}}</p>
        <p>first: {{first_filter|first}}</p>
        <p>last: {{last_filter|last}}</p>
        <p>floatformat: {{floatformat_filter|floatformat}}</p>
        <p>floatformat: {{floatformat_filter|floatformat:2}}</p>
        <p>join: {{join_filter|join:'-'}}</p>
        <p>length: {{length_filter|length}}</p>
        <p>lower: {{lower_filter|lower}}</p>
```

```
        <p>upper: {{upper_filter|upper}}</p>
        <p>random: {{random_filter|random}}</p>
        <p>safe: {{safe_filter|safe}}</p>
        <p>slice: {{slice_filter|slice:'4:10'}}</p>
        <p>striptags: {{striptags_filter|striptags}}</p>
        <p>truncatechars: {{truncatechars_filter|truncatechars:5}}</p>
        <p>truncatechars_html: {{truncatechars_htmls_filter|truncatechars_html:5}}</p>
    </body>
</html>
```

（4）运行上述项目，打开浏览器并在网址栏输入 http://127.0.0.1:8000/，其显示内容如图 5-69 所示。

图 5-69　浏览器中的显示内容

2．自定义过滤器

在实际的开发过程中，内置过滤器往往无法满足一些特殊需求，因此 DTL 还支持自定义过滤器。

自定义过滤器的本质就是自定义一个函数，其创建步骤可以分为 4 步。

(1) 在应用中创建一个 Python 包 templatetags(该包名称不可以进行自定义)。

(2) 首先在 templatetags 包中创建自定义过滤器文件,并在该文件中编写自定义过滤器函数,然后使用 Library 对象的 filter() 方法对自定义过滤器进行注册。

(3) 将当前应用添加到配置文件中的变量 INSTALLED_APPS 之中,完成应用注册操作。

(4) 在模板中使用标签{% load… %}加载自定义过滤器。

下面通过一个示例演示一下如何自定义过滤器。

(1) 创建名为 djangoProject 的 Django 项目。

(2) 打开命令提示符窗口,进入项目所在的根目录中,激活虚拟环境,并输入命令 python manage.py startapp book,创建名为 book 的应用,如图 5-70 所示。

图 5-70 创建应用

(3) 在应用 book 中创建 Python 包 templatetags,如图 5-71 所示。

图 5-71 在应用中创建包 templatetags

(4) 在包 templatetags 中创建文件 my_filter.py,代码如下:

```
#资源包\Code\chapter5\5.5\14\djangoProject\book\templatetags\my_filter.py
from django import template
#自定义过滤器函数,该函数最多只能拥有两个参数
def add_filter(value, word = None):
    return value + '+' + word
#注册自定义过滤器
register = template.Library()
#filter(name, filter_func),其中,参数 name 表示自定义过滤器名称;参数 filter_func 表示自定
#义过滤器函数
register.filter('add_filter', add_filter)
```

(5) 打开应用 book 中的文件 views.py,代码如下:

```
#资源包\Code\chapter5\5.5\14\djangoProject\book\views.py
from django.shortcuts import render
def index(request):
    context = {
        'course': 'Python'
    }
    return render(request, 'index.html', context = context)
```

(6) 打开应用 djangoProject 中的文件 urls.py,代码如下:

```
#资源包\Code\chapter5\5.5\14\djangoProject\djangoProject\urls.py
```

```
from django.urls import path
from book import views
urlpatterns = [
    path('book/', views.index),
]
```

(7) 在项目根目录下的文件夹 templates 中创建模板文件 index.html,代码如下：

```
# 资源包\Code\chapter5\5.5\14\djangoProject\templates\index.html
<!-- 加载自定义过滤器 -->
{% load my_filter %}
<!DOCTYPE html>
<html lang="en">
    <head>
        <meta charset="UTF-8">
        <title>Title</title>
    </head>
    <body>
        自定义过滤器 add_filter: {{ course|add_filter:'Django' }}
    </body>
</html>
```

(8) 打开应用 djangoProject 中的配置文件 settings.py,将应用 book 添加到变量 INSTALLED_APPS 中,完成应用注册,代码如下：

```
# 资源包\Code\chapter5\5.5\14\djangoProject\djangoProject\settings.py
INSTALLED_APPS = [
    'django.contrib.admin',
    'django.contrib.auth',
    'django.contrib.contenttypes',
    'django.contrib.sessions',
    'django.contrib.messages',
    'django.contrib.staticfiles',
    'book',
]
```

(9) 运行上述项目,打开浏览器并在网址栏输入 http://127.0.0.1:8000/book/,其显示内容如图 5-72 所示。

图 5-72　浏览器中的显示内容

5.5.8　模板继承

模板继承是 DTL 的重要特性之一。通过模板继承,可以将模板中重复出现的元素提取出来,并存放在一个已定义的父模板之中,进而达到避免重复编写代码的目的。

模板继承可以分为 3 步。

(1) 在父模板中使用标签{% block…%}…{% endblock %}保存 Web 页面中的常用元素。

(2) 在子模板中使用标签{% extends… %}继承父模板。

(3) 在子模板中使用标签{% block… %}…{% endblock %}将子模板中的内容插入父模板中,需要注意的是,子模板标签{% block…%}…{% endblock %}中的内容会覆盖父模板中的内容。

此外还需要注意两点,一是如果需要保留父模板标签{% block…%}…{% endblock %}中的内容,则需要在子模板中的标签{% block…%}…{% endblock %}内使用标签{{ block.super }};二是子模板中的标签{% block… %}…{% endblock %}之外的代码将不会被模板引擎渲染。

下面通过一个示例演示一下如何使用模板继承。

(1) 创建名为 djangoProject 的 Django 项目。

(2) 打开应用 djangoProject 中的文件 urls.py,代码如下:

```
# 资源包\Code\chapter5\5.5\15\djangoProject\djangoProject\urls.py
from django.urls import path
from django.shortcuts import render
def index(request):
    return render(request, 'index.html')
urlpatterns = [
    path('', index),
]
```

(3) 在项目根目录下的文件夹 templates 中创建模板文件 base.html,代码如下:

```
# 资源包\Code\chapter5\5.5\15\djangoProject\templates\base.html
<!DOCTYPE html>
<html lang="en">
    <head>
        <meta charset="UTF-8">
        <title>{% block title %}老夏学院{% endblock %}</title>
    </head>
    <body>
        {% block head %}
            <h3>网站顶部内容(父模板)</h3>
        {% endblock %}
        {% block main %}
            <h1>网站主体部分(父模板)</h1>
        {% endblock %}
        {% block footer %}
            <h3>网站底部内容(父模板)</h3>
        {% endblock %}
    </body>
</html>
```

(4) 在项目根目录下的文件夹 templates 中创建模板文件 index.html,代码如下:

```
# 资源包\Code\chapter5\5.5\15\djangoProject\templates\index.html
{% extends "base.html" %}
{% block title %}
```

```
    老夏学院-首页
{% endblock %}
{% block main %}
    {{ block.super }}
    <h1 style = "background: greenyellow">网站主题内容(子模板index)</h1>
{% endblock %}
<!-- 此部分内容不会显示 -->
老夏学院: http://www.oldxia.com
```

(5) 运行上述项目,打开浏览器并在网址栏输入 http://127.0.0.1:8000/,其显示内容如图 5-73 所示。

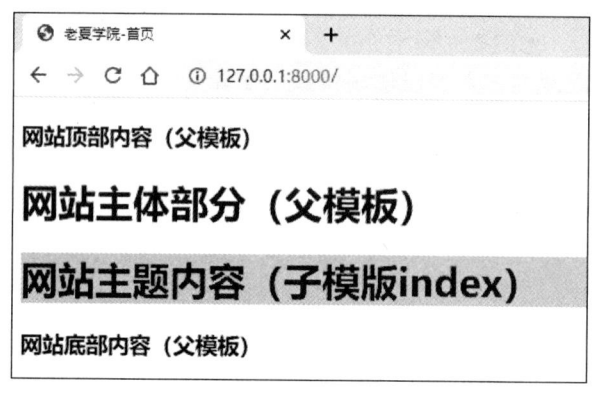

图 5-73　浏览器中的显示内容

5.5.9　加载静态文件

常用的静态文件主要包括 CSS 文件、JavaScript 脚本和图片等。

在 DTL 中,主要通过 static 标签进行静态文件的加载,其步骤可以分为 5 步。

(1) 确保项目同名的应用中的配置文件中的变量 INSTALLED_APPS 包含值 django.contrib.staticfiles。

(2) 确保项目同名的应用中的配置文件中的变量 STATIC_URL 的值不为空,其值为静态文件的访问路径。

(3) 在项目同名的应用中的配置文件中添加变量 STATICFILES_DIRS,其值为存放静态文件的路径,其可以使 DTL 优先在该路径下查找静态文件。

(4) 在项目根目录下,或者在应用所在的目录下创建文件夹 static,用于存放静态文件。

(5) 首先在模板中使用标签{% load… %}加载 static 标签,然后在需要加载静态文件的地方使用标签{% static… %}即可。

此外,如不想每次在模板中加载静态文件时都使用 load 标签,则需要在项目同名的应用中的配置文件中的变量 TEMPLATES 中的 OPTIONS 添加值 'builtins':['django.templatetags.static'],即将 static 标签设置为内置标签。

下面通过一个示例演示一下如何加载静态文件。

(1) 创建名为 djangoProject 的 Django 项目。

(2) 打开命令提示符窗口,进入项目所在的根目录中,激活虚拟环境,并输入命令 python manage.py startapp front 和 python manage.py startapp cms,创建名为 front 和 cms 的应

用,如图 5-74 所示。

```
(django_env) E:\Python全栈开发\djangoProject>python manage.py startapp front
(django_env) E:\Python全栈开发\djangoProject>python manage.py startapp cms
```

图 5-74　创建应用

（3）在项目根目录下创建文件夹 static,以及在应用 front 和 cms 中创建文件夹 templates 和 static,如图 5-75 所示。

图 5-75　创建文件夹

（4）打开应用 djangoProject 中的配置文件 settings.py,完成以下 3 个操作,一是将应用 front 和 cms 添加到变量 INSTALLED_APPS 中;二是添加变量 STATICFILES_DIRS;三是给变量 TEMPLATES 中的 OPTIONS 添加值,代码如下:

```
#资源包\Code\chapter5\5.5\16\djangoProject\djangoProject\settings.py
INSTALLED_APPS = [
    'django.contrib.admin',
    'django.contrib.auth',
    'django.contrib.contenttypes',
    'django.contrib.sessions',
    'django.contrib.messages',
    'django.contrib.staticfiles',
    'front',
    'cms',
]
TEMPLATES = [
```

```
        {
            'BACKEND': 'django.template.backends.django.DjangoTemplates',
            'DIRS': [BASE_DIR / 'templates']
,
            'APP_DIRS': True,
            'OPTIONS': {
                'context_processors': [
                    'django.template.context_processors.debug',
                    'django.template.context_processors.request',
                    'django.contrib.auth.context_processors.auth',
                    'django.contrib.messages.context_processors.messages',
                ],
                'builtins':['django.templatetags.static']
            },
        },
]
STATICFILES_DIRS = [
    os.path.join(BASE_DIR,"static")
]
```

（5）打开应用 front 中的文件 views.py，代码如下：

```
# 资源包\Code\chapter5\5.5\16\djangoProject\front\views.py
from django.shortcuts import render
def front(request):
    return render(request, 'front_index.html')
```

（6）打开应用 cms 中的文件 views.py，代码如下：

```
# 资源包\Code\chapter5\5.5\16\djangoProject\cms\views.py
from django.shortcuts import render
def cms(request):
    return render(request, 'cms_index.html')
```

（7）打开应用 djangoProject 中的文件 urls.py，代码如下：

```
# 资源包\Code\chapter5\5.5\16\djangoProject\djangoProject\urls.py
from django.urls import path
from django.shortcuts import render
from front import views as front_views
from cms import views as cms_views
def index(request):
    return render(request, 'index.html')
urlpatterns = [
    path('', index),
    path('front/', front_views.front),
    path('cms/', cms_views.cms),
]
```

（8）在项目根目录下的文件夹 static 中创建文件 index.css，代码如下：

```
# 资源包\Code\chapter5\5.5\16\djangoProject\static\index.css
h1{
    font-size: 30px;
    font-weight: bold;
```

```
    color: red;
}
```

(9) 为了避免加载时产生混淆，在应用 front 的文件夹 static 中创建文件夹 front，并在其中创建文件 index.css，代码如下：

```
#资源包\Code\chapter5\5.5\16\djangoProject\front\static\front\index.css
h1{
    font-size: 25px;
    font-weight: bold;
color: blue;
}
```

(10) 在应用 front 的文件夹 static 中的文件夹 front 内创建文件 index.js，代码如下：

```
#资源包\Code\chapter5\5.5\16\djangoProject\front\static\front\index.js
alert("欢迎访问 front 页面!")
```

(11) 为了避免加载时产生混淆，在应用 cms 的文件夹 static 中创建文件夹 cms，并在其中创建文件 index.css，代码如下：

```
#资源包\Code\chapter5\5.5\16\djangoProject\cms\static\cms\index.css
h1{
    font-size: 25px;
    font-weight: bold;
    color: green;
}
```

(12) 在应用 cms 的文件夹 static 中的文件夹 cms 内创建文件 index.js，代码如下：

```
#资源包\Code\chapter5\5.5\16\djangoProject\cms\static\cms\index.js
alert("欢迎访问 cms 页面!")
```

(13) 在项目根目录下的文件夹 templates 中创建模板文件 index.html，代码如下：

```
#资源包\Code\chapter5\5.5\16\djangoProject\templates\index.html
<!DOCTYPE html>
<html lang="en">
    <head>
        <meta charset="UTF-8">
        <title>Title</title>
        <link rel="stylesheet" href="{% static 'index.css' %}">
    </head>
    <body>
        <h1>这是 index 页面</h1>
    </body>
</html>
```

(14) 在应用 front 的文件夹 templates 中创建模板文件 front_index.html，代码如下：

```
#资源包\Code\chapter5\5.5\16\djangoProject\front\templates\front_index.html
<!DOCTYPE html>
<html lang="en">
```

```html
    <head>
        <meta charset="UTF-8">
        <title>Title</title>
        <link rel="stylesheet" href="{% static 'front/index.css' %}">
        <script src="{% static 'front/index.js' %}"></script>
    </head>
    <body>
        <h1>这是 front 页面</h1>
    </body>
</html>
```

(15) 在应用 cms 的文件夹 templates 中创建模板文件 cms_index.html，代码如下：

```html
#资源包\Code\chapter5\5.5\16\djangoProject\cms\templates\cms_index.html
<!DOCTYPE html>
<html lang="en">
    <head>
        <meta charset="UTF-8">
        <title>Title</title>
        <link rel="stylesheet" href="{% static 'cms/index.css' %}">
        <script src="{% static 'cms/index.js' %}"></script>
    </head>
    <body>
        <h1>这是 cms 页面</h1>
    </body>
</html>
```

(16) 运行上述项目，打开浏览器并在网址栏输入 http://127.0.0.1:8000/，其显示内容如图 5-76 所示。

图 5-76　浏览器中的显示内容(1)

(17) 在浏览器中的网址栏输入 http://127.0.0.1:8000/front/，其显示内容如图 5-77 所示。

图 5-77　浏览器中的显示内容(2)

(18) 在浏览器中的网址栏输入 http://127.0.0.1:8000/cms/，其显示内容如图 5-78 所示。

图 5-78　浏览器中的显示内容(3)

5.6　类视图

除视图函数之外，视图也可以基于类来实现，而类视图的好处就是支持继承，即首先可将共性的内容抽取出来放到父类中，然后在子类中完成各自的业务逻辑，并继承父类即可。

类视图是一个 Python 类，它继承自 django.views.generic 中的 View 类，并实现了 View 类中定义的方法。

类视图中具有 get()、post()、put()、delete()和 head()等方法，用于处理不同的 HTTP 请求，而当客户端发起的 HTTP 请求中使用了视图不支持的方法时，可以通过 http_method_not_allowed()方法进行处理。

此外，在配置类视图的路由时，不能直接传入该类的名称，而是需要使用类视图的 as_view()方法将类转换成可以为路由注册的视图函数。

下面通过一个示例演示一下如何使用类视图。

(1) 创建名为 djangoProject 的 Django 项目。

(2) 打开命令提示符，进入项目所在的根目录中，激活虚拟环境，并输入命令 python manage.py startapp book，创建名为 book 的应用，如图 5-79 所示。

图 5-79　创建应用

(3) 在项目根目录下的文件夹 templates 中创建模板文件 index.html，代码如下：

```html
# 资源包\Code\chapter5\5.6\1\djangoProject\templates\index.html
<html lang = "en">
    <head>
        <meta charset = "UTF-8">
        <title>Title</title>
    </head>
    <body>
        <form action = "" method = "post">
            {% csrf_token %}
            <p>作者名称：<input type = "text" name = "book_author"></p>
            <input type = "submit" value = "提交">
        </form>
    </body>
</html>
```

(4) 打开应用 book 中的文件 views.py，代码如下：

```
# 资源包\Code\chapter5\5.6\1\djangoProject\book\views.py
from django.http import HttpResponse
from django.shortcuts import render
from django.views.generic import View
class BookDetailView(View):
    def get(self, request):
        return render(request, "index.html")
    def post(self, request):
        book_author = request.POST.get('book_author')
        return HttpResponse(f'作者姓名：{book_author}')
class BookListView(View):
    def post(self, request):
        return HttpResponse('这是 POST 请求')
    def http_method_not_allowed(self, request):
        return HttpResponse('不支持非 POST 请求的请求方法！')
```

（5）打开应用 djangoProject 中的文件 urls.py，代码如下：

```
# 资源包\Code\chapter5\5.6\1\djangoProject\djangoProject\urls.py
from django.urls import path
from book import views
urlpatterns = [
    path('bookdetail/', views.BookDetailView.as_view()),
    path('booklist/', views.BookListView.as_view()),
]
```

（6）运行上述项目，打开浏览器并在网址栏输入 http://127.0.0.1:8000/bookdetail/，其显示内容如图 5-80 所示。

此时，输入作者名称"夏正东"，并单击"提交"按钮，其显示内容如图 5-81 所示。

图 5-80　浏览器中的显示内容(1)

图 5-81　浏览器中的显示内容(2)

图 5-82　浏览器中的显示内容(3)

（7）在浏览器中的网址栏输入 http://127.0.0.1:8000/booklist/，其显示内容如图 5-82 所示。

此外，在类视图中可以通过 django.utils.decorators 模块中的 method_decorator()方法将函数装饰器应用于类视图中的方法或类视图本身。

下面通过一个示例，演示一下如何给类视图添加装饰器。

（1）创建名为 djangoProject 的 Django 项目。

（2）打开命令提示符窗口，进入项目所在的根目录中，激活虚拟环境，并输入命令 python manage.py startapp book，创建名为 book 的应用，如图 5-83 所示。

```
E:\Python全栈开发\djangoProject>workon django_env
(django_env) E:\Python全栈开发\djangoProject>python manage.py startapp book
```

图 5-83 创建应用

(3) 打开应用 book 中的文件 views.py，代码如下：

```python
# 资源包\Code\chapter5\5.6\2\djangoProject\book\views.py
from django.views.generic import View
from django.shortcuts import redirect, reverse
from django.http import HttpResponse
from django.utils.decorators import method_decorator
def login_required(func):
    def wrapper(request):
        username = request.GET.get('username')
        if username:
            return func(request, username)
        else:
            return redirect(reverse('login'))
    return wrapper
# 装饰类视图
@method_decorator(login_required, name='get')
class IndexView(View):
    # 装饰类视图中的方法
    # @method_decorator(login_required)
    def get(self, request, username):
        return HttpResponse(f'这是登录之后的页面,欢迎用户{username}!')
def login(request):
    return HttpResponse('登录页面')
```

(4) 打开应用 djangoProject 中的文件 urls.py，代码如下：

```python
# 资源包\Code\chapter5\5.6\2\djangoProject\djangoProject\urls.py
from django.urls import path
from book import views
urlpatterns = [
    path('', views.IndexView.as_view()),
    path('login/', views.login, name='login'),
]
```

(5) 运行上述项目，打开浏览器并在网址栏输入 http://127.0.0.1:8000/login/，其显示内容如图 5-84 所示。

(6) 在浏览器中的网址栏输入 http://127.0.0.1:8000/?username=oldxia，其显示内容如图 5-85 所示。

图 5-84 浏览器中的显示内容(1)　　图 5-85 浏览器中的显示内容(2)

5.7 数据库

在 Django 中，可以通过 ORM 技术对数据库进行访问，其允许开发者使用 Python 对象来操作数据库，而不必直接编写 SQL 查询。

ORM技术将数据库中的表映射为一个Python类,将表中的一条数据映射为类的实例,将表的标签映射为类的属性,这使数据库的操作更容易、更直观。

5.7.1 定义数据模型

创建数据模型可以分为5步。

(1) 在项目同名的应用中的配置文件内,通过修改变量DATABASES的值,用于进行数据库的相关配置,以便达到连接数据库的目的。

(2) 在应用中的文件models.py中定义数据模型。

(3) 将定义数据模型的应用添加至项目同名的应用中的配置文件内的变量INSTALLED_APP。

(4) 打开命令提示符窗口,进入项目所在的目录中,激活虚拟环境,并执行命令python manage.py makemigrations,用于生成迁移脚本文件。

(5) 执行命令python manage.py migrate,用于将迁移的脚本文件映射到数据库中。

下面通过一个示例演示一下如何定义数据模型。

(1) 创建名为djangoProject的Django项目。

(2) 打开命令提示符窗口,进入项目所在的根目录中,激活虚拟环境,并输入命令python manage.py startapp book,创建名为book的应用,如图5-86所示。

```
E:\Python全栈开发\djangoProject>workon django_env
(django_env) E:\Python全栈开发\djangoProject>python manage.py startapp book
```

图5-86 创建应用

(3) 打开应用djangoProject中的文件__init__.py,代码如下:

```
# 资源包\Code\chapter5\5.7\1\djangoProject\djangoProject\__init__.py
import pymysql
pymysql.install_as_Mysqldb()
```

(4) 打开应用djangoProject中的配置文件settings.py,进行数据库的相关配置,用于连接数据库,代码如下:

```
# 资源包\Code\chapter5\5.7\1\djangoProject\djangoProject\settings.py
DATABASES = {
    'default': {
        # 数据库引擎
        'ENGINE': 'django.db.backends.mysql',
        # 数据库名称
        'NAME': 'django_db',
        # 连接MySQL的用户名
        'USER': 'root',
        # 连接MySQL的密码
        'PASSWORD': '12345678',
        # MySQL的主机地址
        'HOST': '127.0.0.1',
        # MySQL的端口号
        'PORT': '3306',
    }
}
```

（5）打开应用 djangoProject 中的配置文件 settings.py，将应用 book 添加到变量 INSTALLED_APPS 中，用于完成应用注册，代码如下：

```
# 资源包\Code\chapter5\5.7\1\djangoProject\djangoProject\settings.py
INSTALLED_APPS = [
    'django.contrib.admin',
    'django.contrib.auth',
    'django.contrib.contenttypes',
    'django.contrib.sessions',
    'django.contrib.messages',
    'django.contrib.staticfiles',
    'book',
]
```

（6）打开应用 book 中的文件 models.py，用于定义数据模型，代码如下：

```
# 资源包\Code\chapter5\5.7\1\djangoProject\book\models.py
from django.db import models
class Book(models.Model):
    id = models.AutoField(primary_key=True)
    name = models.CharField(max_length=20, null=False)
    author = models.CharField(max_length=20, null=False)
    price = models.FloatField(null=False, default=0)
```

（7）打开命令提示符窗口，登入 MySQL。

（8）在 MySQL 命令行窗口中输入 SQL 语句"create database django_db;"，创建数据库 django_db，其结果如图 5-87 所示。

（9）打开命令提示符，进入项目所在的根目录中，激活虚拟环境，并输入命令 python manage.py makemigrations，生成迁移脚本文件，如图 5-88 所示。

图 5-87　创建数据库　　　　图 5-88　生成迁移脚本文件

（10）再次输入命令 python manage.py migrate，将迁移脚本文件映射到数据库中，如图 5-89 所示。

（11）在 MySQL 命令行窗口中输入 SQL 语句"show tables;"，查询当前数据库中的数据表，其结果如图 5-90 所示。

图 5-89　将迁移脚本文件映射到数据库中　　　　图 5-90　查询数据表

其中,数据表 book_book 就是应用 book 中 Book 类映射到数据库中的数据表。

(12) 再次输入 SQL 语句"desc book_book;",查询该数据表的表结构,其结果如图 5-91 所示。

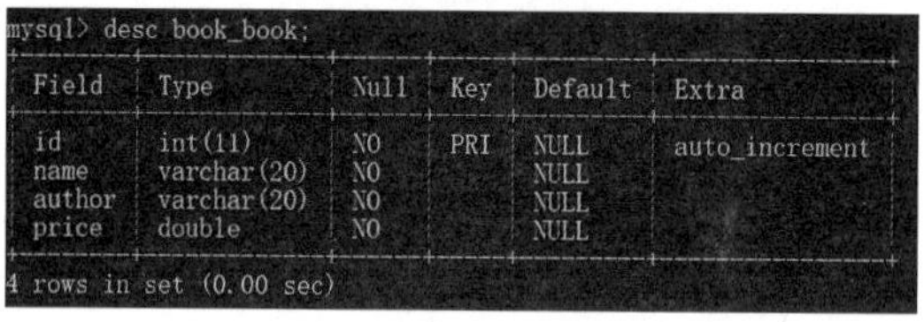

```
mysql> desc book_book;
+--------+-------------+------+-----+---------+----------------+
| Field  | Type        | Null | Key | Default | Extra          |
+--------+-------------+------+-----+---------+----------------+
| id     | int(11)     | NO   | PRI | NULL    | auto_increment |
| name   | varchar(20) | NO   |     | NULL    |                |
| author | varchar(20) | NO   |     | NULL    |                |
| price  | double      | NO   |     | NULL    |                |
+--------+-------------+------+-----+---------+----------------+
4 rows in set (0.00 sec)
```

图 5-91 数据表 book_book 中的表结构

5.7.2 Manager 类和 QuerySet 类

在 Django 中,每个数据模型都有一个默认的 Manager 类(管理器类)的实例对象 objects,用于提供对数据库中相应表的操作接口,而 QuerySet 类则用于数据库的所有查询及更新交互,其允许开发人员以面向对象的方式操作数据库,并且其提供了丰富的方法和链式操作,使查询数据库变得非常方便和灵活。

此外,Manager 类本身不具有任何属性和方法,其方法均是通过 Python 动态添加的方式从 QuerySet 类中复制过来的,所以 Manager 类的绝大部分方法基于 QuerySet 类,并且一个 QuerySet 类可以包含一个或多个数据模型对象。

QuerySet 类的常用方法如表 5-7 所示。

表 5-7 QuerySet 类的常用方法

方法	描述
filter()	返回满足指定条件的数据
exclude()	返回不满足指定条件的数据
get()	返回满足指定条件的唯一数据
all()	返回数组模型中的所有数据
first()	返回第 1 条数据
last()	返回最后一条数据
distinct()	返回去重后的数据
annotate()	返回使用聚合函数进行操作后的数据,支持分组聚合
aggregate()	返回使用聚合函数进行操作后的数据
order_by()	返回按照指定字段进行排序的数据
reverse()	返回反向排序的数据
values()	返回给定字段的数据所组成的字典
values_list()	返回给定字段的数据所组成的元组
select_related()	根据外键关系(一对多或一对一)返回相关联数据模型中的数据
prefetch_related()	根据外键关系(多对多或多对一)返回相关联数据模型中的数据
defer()	返回除指定字段之外的所有数据
only()	返回指定字段所对应的数据
create()	创建一条数据
get_or_create()	创建或返回一条数据
bulk_create()	创建多条数据

续表

方法	描述
count()	返回数据的数量
exists()	判断指定条件的数据是否存在
update()	根据指定条件更新数据
delete()	根据指定条件删除数据

除此之外,如果想获取部分数据,则可以对 QuerySet 对象使用切片操作,即相当于在数据库层面使用 LIMIT 和 OFFSET 操作。

5.7.3 查询条件

查询条件作为 QuerySet 类中方法的参数,用于在 Django 框架中进行数据库查询时对结果进行过滤的一种机制,其允许开发者根据特定的条件来筛选出符合要求的数据。

查询条件可以分为 4 大类,即查询过滤器、聚合函数、F 表达式和 Q 表达式。

1. 查询过滤器

查询过滤器通过"字段＋__＋查询过滤器"的方式实现,其常用的查询过滤器如表 5-8 所示。

表 5-8 常用的查询过滤器

查询过滤器	描述
exact	根据指定字段的值进行精确匹配查询
iexact	根据指定字段的值进行精确匹配查询(不区分大小写)
contains	筛选出根据指定字段包含指定值的数据
icontains	筛选出根据指定字段包含指定值的数据(不区分大小写)
in	筛选出指定字段的值在给定列表中的数据
gt	筛选出指定字段的值大于指定值的数据
gte	筛选出指定字段的值大于或等于指定值的数据
lt	筛选出指定字段的值小于指定值的数据
lte	筛选出指定字段的值小于或等于指定值的数据
startswith	筛选出指定字段的值以指定值开头的数据
endswith	筛选出指定字段的值以指定值结尾的数据
range	筛选出指定字段的值在给定范围内的数据
date	筛选出指定字段的值为指定日期的数据
time	筛选出指定字段的值为指定时间的数据
year	筛选出指定字段的值为指定年份的数据
week_day	筛选出指定字段的值为指定星期的数据
isnull	筛选出指定字段的值为空值的数据
regex	根据指定字段的值进行正则表达式匹配

2. 聚合函数

在 Django 中,聚合函数通过 aggregate()方法或 annotate()方法实现,其常用的聚合函数如表 5-9 所示。

表 5-9 常用的聚合函数

聚合函数	描述
Count()	统计指定字段的个数
Sum()	求和
Max()	求最大值
Min()	求最小值
Avg()	求平均值

3. F 表达式

F 表达式是一个强大的工具，允许在查询表达式中引用模型的字段，从而在数据库层面执行各种操作，而无须将数据加载到 Python 内存中。这不仅可以提高性能，还允许利用数据库的优化功能，尤其是在处理大量数据时，可以显著地提高性能并减少资源消耗。

F 表达式的常用应用场景：字段引用，即在查询表达式中直接引用模型的字段；算术运算，即使用 F() 函数执行算术运算，如加、减、乘、除等；条件表达式，F() 函数可用于创建复杂的条件逻辑；更新字段值，即在更新操作中，F() 函数可以用于基于现有字段值的更新；数据库函数调用，F() 函数可以与数据库特定的函数结合使用；字段间运算，F() 函数可以用于字段间的运算。

4. Q 表达式

Q 表达式是 Django 框架中用于构建复杂查询表达式的一种功能，其可以灵活地构建查询条件，从而实现更复杂的查询需求。

Q 表达式的应用场景非常广泛，例如在需要筛选多个条件的数据时，或者需要构建复杂的查询逻辑时。

下面通过一个示例演示一下如何进行数据的增、删、改、查等操作。

（1）创建名为 djangoProject 的 Django 项目。

（2）打开命令提示符窗口，进入项目所在的根目录中，激活虚拟环境，并输入命令 python manage.py startapp book，创建名为 book 的应用，如图 5-92 所示。

图 5-92 创建应用

（3）打开应用 djangoProject 中的文件 __init__.py，代码如下：

```
# 资源包\Code\chapter5\5.7\2\djangoProject\djangoProject\__init__.py
import pymysql
pymysql.install_as_Mysqldb()
```

（4）打开应用 djangoProject 中的配置文件 settings.py，进行数据库的相关配置，用于连接数据库，代码如下：

```
# 资源包\Code\chapter5\5.7\2\djangoProject\djangoProject\settings.py
DATABASES = {
    'default': {
        # 数据库引擎
        'ENGINE': 'django.db.backends.mysql',
        # 数据库名称
```

```
        'NAME': 'django_db',
        #连接MySQL的用户名
        'USER': 'root',
        #连接MySQL的密码
        'PASSWORD': '12345678',
        #MySQL的主机地址
        'HOST': '127.0.0.1',
        #MySQL的端口号
        'PORT': '3306',
    }
}
```

(5) 打开应用 djangoProject 中的配置文件 settings.py，将应用 book 添加到变量 INSTALLED_APPS 中，用于完成应用注册，代码如下：

```
#资源包\Code\chapter5\5.7\2\djangoProject\djangoProject\settings.py
INSTALLED_APPS = [
    'django.contrib.admin',
    'django.contrib.auth',
    'django.contrib.contenttypes',
    'django.contrib.sessions',
    'django.contrib.messages',
    'django.contrib.staticfiles',
    'book',
]
```

(6) 打开应用 book 中的文件 models.py，用于定义数据模型，代码如下：

```
#资源包\Code\chapter5\5.7\2\djangoProject\book\models.py
from django.db import models
class Book(models.Model):
    name = models.CharField(max_length = 100)
    author = models.CharField(max_length = 20, null = False)
    price = models.FloatField(null = False, default = 0)
    publishing_time = models.DateTimeField(null = False)
```

(7) 打开应用 book 中的文件 views.py，代码如下：

```
#资源包\Code\chapter5\5.7\2\djangoProject\book\views.py
from book.models import Book
from django.http import HttpResponse
from datetime import datetime
from django.db.models import Avg, Count, Sum, Max, Min, F, Q
from django.db import connection
def add_one_data(request):
    Book.objects.create(name = '《Python全栈开发——Web编程》', author = '夏正东', price = '999',
publishing_time = datetime(year = 2024, month = 12, day = 1))
    return HttpResponse('已成功添加一册图书！')
def add_more_data(request):
    Book.objects.bulk_create([Book(name = '《Python全栈开发——基础入门》', author = '夏正东',
price = 79, publishing_time = datetime(year = 2022, month = 7, day = 1)),
                             Book(name = '《Python全栈开发——高阶编程》', author = '夏正东',
price = 89, publishing_time = datetime(year = 2022, month = 8, day = 1)),
```

```python
                                Book(name = '《Python 全栈开发——数据分析》', author = '夏正东',
price = 79, publishing_time = datetime(year = 2023, month = 2, day = 1))])
    return HttpResponse('已成功添加多册图书!')
def filter_exact(request):
    # 在精确匹配查询时,id__exact = 1 等价于 id = 1
    books = Book.objects.filter(id__exact = 1)
    book = Book.objects.get(id = 1)
    return HttpResponse(f'查询过滤器 exact 等价的 SQL 语句为{books.query}<br>查询结果为
{book.name}')
def filter_contains(request):
    books = Book.objects.filter(name__contains = "Python")
    books_lt = []
    for book in books:
        books_lt.append(book.name)
    return HttpResponse(f'查询过滤器 contains 等价的 SQL 语句为{books.query}<br>查询结果为
{books_lt}')
def filter_gt(request):
    books = Book.objects.filter(price__gt = 100)
    books_lt = []
    for book in books:
        books_lt.append(book.name)
    return HttpResponse(f'查询过滤器 gt 等价的 SQL 语句为{books.query}<br>查询结果为{books_
lt}')
def filter_endswith(request):
    books = Book.objects.filter(name__endswith = "编程》")
    books_lt = []
    for book in books:
        books_lt.append(book.name)
    return HttpResponse(f'查询过滤器 endwith 等价的 SQL 语句为{books.query}<br>查询结果为
{books_lt}')
def filter_range(request):
    books = Book.objects.filter(price__range = (50, 100))
    books_lt = []
    for book in books:
        books_lt.append(book.name)
    return HttpResponse(f'查询过滤器 range 等价的 SQL 语句为{books.query}<br>查询结果为
{books_lt}')
def filter_date(request):
    books = Book.objects.filter(publishing_time__date = datetime(year = 2023, month = 2, day =
1))
    books_lt = []
    for book in books:
        books_lt.append(book.name)
    return HttpResponse(f'查询过滤器 date 等价的 SQL 语句为{books.query}<br>查询结果为
{books_lt}')
def filter_isnull(request):
    books = Book.objects.filter(publishing_time__isnull = True)
    books_lt = []
    for book in books:
        books_lt.append(book.name)
    return HttpResponse(f'查询过滤器 isnull 等价的 SQL 语句为{books.query}<br>查询结果为
{books_lt}')
def filter_regex(request):
    books = Book.objects.filter(name__regex = r"^《")
```

```python
        books_lt = []
        for book in books:
            books_lt.append(book.name)
        return HttpResponse(f'查询过滤器 regex 等价的 SQL 语句为{books.query}<br>查询结果为{books_lt}')
    def aggregate_avg(request):
        books = Book.objects.aggregate(price_avg=Avg("price"))
        return HttpResponse(f'聚合函数 Avg 等价的 SQL 语句为{connection.queries[-1]}<br>查询结果为{books}')
    def aggregate_count(request):
        books = Book.objects.aggregate(book_count=Count("id"))
        return HttpResponse(f'聚合函数 Count 等价的 SQL 语句为{connection.queries[-1]}<br>查询结果为{books}')
    def aggregate_max_min(request):
        books = Book.objects.aggregate(price_max=Max("price"), price_min=Min("price"))
        return HttpResponse(f'聚合函数 Max 和 Min 等价的 SQL 语句为{connection.queries[-1]}<br>查询结果为{books}')
    def aggregate_sum(request):
        books = Book.objects.aggregate(price_sum=Sum("price"))
        return HttpResponse(f'聚合函数 Sum 等价的 SQL 语句为{connection.queries[-1]}<br>查询结果为{books}')
    def expression_f(request):
        Book.objects.update(price=F("price") + 5)
        books = Book.objects.all().order_by('price')
        books_lt = []
        for book in books:
            books_lt.append(book.price)
        return HttpResponse(f'F 表达式等价的 SQL 语句为{connection.queries[-2]}<br>查询结果为{books_lt}')
    def expression_q(request):
        books = Book.objects.filter(Q(price__lte=90) | Q(name__contains="Web"))
        books_lt = []
        for book in books:
            books_lt.append(book.name)
        return HttpResponse(f'Q 表达式等价的 SQL 语句为{connection.queries[-1]}<br>查询结果为{books_lt}')
    def del_data(request):
        Book.objects.filter(name='《Python全栈开发——基础入门》').delete()
        return HttpResponse('删除书籍成功!')
```

(8) 打开应用 djangoProject 中的文件 urls.py，代码如下：

```python
#资源包\Code\chapter5\5.7\2\djangoProject\djangoProject\urls.py
from django.urls import path
from book import views
urlpatterns = [
    path('add_more_data/', views.add_more_data),
    path('add_one_data/', views.add_one_data),
    path('filter_exact/', views.filter_exact),
    path('filter_contains/', views.filter_contains),
    path('filter_gt/', views.filter_gt),
    path('filter_endswith/', views.filter_endswith),
    path('filter_range/', views.filter_range),
    path('filter_date/', views.filter_date),
    path('filter_isnull/', views.filter_isnull),
```

```
    path('filter_regex/', views.filter_regex),
    path('aggregate_avg/', views.aggregate_avg),
    path('aggregate_count/', views.aggregate_count),
    path('aggregate_max_min/', views.aggregate_max_min),
    path('aggregate_sum/', views.aggregate_sum),
    path('expression_f/', views.expression_f),
    path('expression_q/', views.expression_q),
    path('del_data/', views.del_data),
]
```

（9）打开命令提示符窗口，登入 MySQL。

（10）在 MySQL 命令行窗口中输入 SQL 语句"create database django_db;"，创建数据库 django_db，其结果如图 5-93 所示。

（11）打开命令提示符窗口，进入项目所在的根目录中，激活虚拟环境，并输入命令 python manage.py makemigrations，生成迁移脚本文件，如图 5-94 所示。

图 5-93　创建数据库

图 5-94　生成迁移脚本文件

（12）再次输入命令 python manage.py migrate，将迁移脚本文件映射到数据库中，如图 5-95 所示。

图 5-95　将迁移脚本文件映射到数据库中

（13）运行上述项目，打开浏览器并在网址栏输入 http://127.0.0.1:8000/add_more_data/，其显示内容如图 5-96 所示。

（14）在浏览器中的网址栏输入 http://127.0.0.1:8000/add_one_data/，其显示内容如图 5-97 所示。

图 5-96　浏览器中的显示内容（1）

图 5-97　浏览器中的显示内容（2）

(15) 打开命令提示符窗口，登入 MySQL。

(16) 在 MySQL 命令行窗口中输入 SQL 语句"select * from book_book;"，其结果如图 5-98 所示。

图 5-98　数据表 book_book 中的数据

(17) 在浏览器中的网址栏输入 http://127.0.0.1:8000/filter_exact/，其显示内容如图 5-99 所示。

图 5-99　浏览器中的显示内容(1)

(18) 在浏览器中的网址栏输入 http://127.0.0.1:8000/filter_contains/，其显示内容如图 5-100 所示。

图 5-100　浏览器中的显示内容(2)

(19) 在浏览器中的网址栏输入 http://127.0.0.1:8000/filter_gt/，其显示内容如图 5-101 所示。

图 5-101　浏览器中的显示内容(3)

(20) 在浏览器中的网址栏输入 http://127.0.0.1:8000/filter_endswith/，其显示内容如图 5-102 所示。

图 5-102　浏览器中的显示内容(4)

(21) 在浏览器中的网址栏输入 http://127.0.0.1:8000/filter_range/，其显示内容如图 5-103 所示。

图 5-103　浏览器中的显示内容(5)

(22)在浏览器中的网址栏输入 http://127.0.0.1:8000/filter_date/，其显示内容如图 5-104 所示。

图 5-104　浏览器中的显示内容(6)

(23)在浏览器中的网址栏输入 http://127.0.0.1:8000/filter_isnull/，其显示内容如图 5-105 所示。

图 5-105　浏览器中的显示内容(7)

(24)在浏览器中的网址栏输入 http://127.0.0.1:8000/filter_regex/，其显示内容如图 5-106 所示。

图 5-106　浏览器中的显示内容(8)

(25)在浏览器中的网址栏输入 http://127.0.0.1:8000/aggregate_avg/，其显示内容如图 5-107 所示。

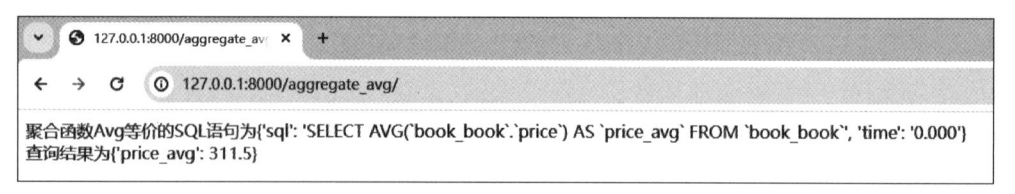

图 5-107　浏览器中的显示内容(9)

(26)在浏览器中的网址栏输入 http://127.0.0.1:8000/aggregate_count/，其显示内容如图 5-108 所示。

图 5-108　浏览器中的显示内容(10)

(27)在浏览器中的网址栏输入 http://127.0.0.1:8000/aggregate_max_min/，其显示内容如图 5-109 所示。

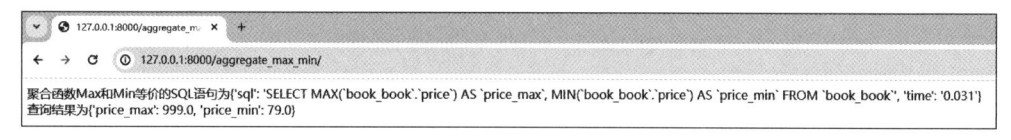

图 5-109　浏览器中的显示内容(11)

（28）在浏览器中的网址栏输入 http://127.0.0.1:8000/aggregate_sum/，其显示内容如图 5-110 所示。

图 5-110　浏览器中的显示内容（12）

（29）在浏览器中的网址栏输入 http://127.0.0.1:8000/expression_q/，其显示内容如图 5-111 所示。

图 5-111　浏览器中的显示内容（13）

（30）在浏览器中的网址栏输入 http://127.0.0.1:8000/expression_f/，其显示内容如图 5-112 所示。

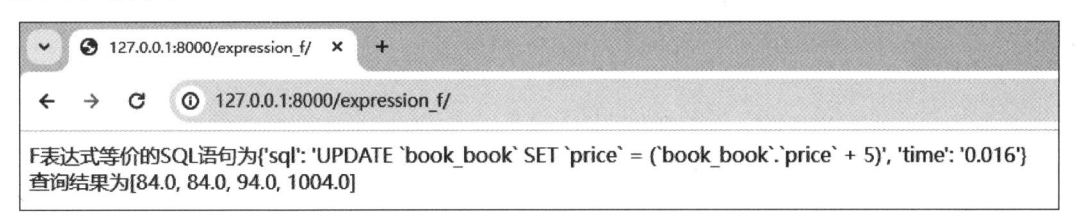

图 5-112　浏览器中的显示内容（14）

（31）在浏览器中的网址栏输入 http://127.0.0.1:8000/del_data/，其显示内容如图 5-113 所示。

图 5-113　浏览器中的显示内容（15）

（32）在 MySQL 命令行窗口中输入 SQL 语句"select * from book_book;"，其结果如图 5-114 所示。

图 5-114　数据表 book_book 中的数据

5.7.4 常用字段

在 ORM 中，Python 类中的属性对应为数据表中的字段，而字段中的属性则可以通过各种字段类型进行创建，其常用的字段类型如表 5-10 所示。

表 5-10 常用的字段类型

字 段 类 型	字段类型常用属性	描 述
AutoField()	参数 null 用于设置在数据库中字段是否为空。参数 blank 用于设置在表单验证时字段是否为空。参数 db_column 用于设置字段是否为空。参数 default 用于设置字段的默认值。参数 primary_key 用于设置字段是否为主键。参数 unique 用于设置字段的值是否唯一，即不允许有重复值	自动增长类型字段
BooleanField()		布尔类型字段
NullBooleanField()		允许为空的布尔类型字段
CharField()		字符串类型字段
DataField()		日期类型字段
DateTImeField()		日期和时间类型字段
TimeField()		时间类型字段
EmailField()		提供验证 Email 机制的字符串类型字段
ImageField()		图像类型字段
FileField()		文件类型字段
FloatField()		浮点数类型字段
IntegerField()		整数类型字段
BigIntegerField()		长整型字段
PositiveIntegerField()		正整型字段
SmallIntegerField()		短整型字段
PositiveSmallIntegerField()		正短整型字段
TextField()		文本类型字段
UUIDField()		提供验证 UUID 机制的字符串类型字段
URLField()		提供验证 URL 机制的字符串类型字段
ForeignKey		外键类型字段

5.7.5 Meta 类

每个数据模型类中都有一个子类 Meta，该类中封装了一些数据库信息，称为数据模型的元数据。

Django 会将 Meta 类中的元数据选项定义附加到数据模型中，常见的元数据定义有 db_table（数据表名称）、abstract（抽象类）和 ordering（字段排序）等，而 Meta 类作为内部类，它定义的元数据可以让 admin 管理后台更加友好，并且数据的可读性更高。

此外，由于 Meta 类定义的元数据相当于数据模型的配置信息，所以开发人员可以根据自己的需求选择性地进行添加，而当没有需求时也可以不定义 Meta 类，此时，Django 会应用默认的元数据。

5.7.6 外键

在 MySQL 中，外键可以让多表之间的关系更加紧密，而 Django 同样支持外键，其通过数据模型中的 ForeignKey 字段实现，其语法格式如下：

```
ForeignKey(to, on_delete, related_name, related_query_name)
```

其中,参数 to 表示关系另一侧的数据模型名称,参数 on_delete 用于设置外键约束,其包括以下 5 种,即 models.CASCADE(表示级联删除,即删除父表的某一条数据时,其子表中使用该外键的数据也会被删除)、models.PROTECT(表示保护模式,即删除父表的某一条数据时会抛出 ProtectedError 异常,阻止删除操作)、models.SET_NULL(表示设置为 NULL,即删除父表的某一条数据时,其子表中的外键会被设置为 NULL)、models.SET_DEFAULT(表示设置为默认值,即删除父表的某一条数据时,其子表中的外键会被设置为默认值)和 models.SET(表示设置为指定值,即删除父表的某一条数据时,其子表中的外键会被设置为自定义的外键字段的值),参数 related_name 用于指定反向引用的名称;参数 related_query_name 用于在关联数据模型查询时,指定字段名称,默认使用的是数据模型的名称。

下面通过一个示例演示一下如何使用外键。

(1)创建名为 djangoProject 的 Django 项目。

(2)打开命令提示符窗口,进入项目所在的根目录中,激活虚拟环境,并输入命令 python manage.py startapp book 和 python manage.py startapp press,创建名为 book 和 press 的应用,如图 5-115 所示。

```
(django_env) E:\Python全栈开发\djangoProject>python manage.py startapp book
(django_env) E:\Python全栈开发\djangoProject>python manage.py startapp press
```

图 5-115　创建应用

(3)打开应用 djangoProject 中的文件 __init__.py,代码如下:

```python
# 资源包\Code\chapter5\5.7\3\djangoProject\djangoProject\__init__.py
import pymysql
pymysql.install_as_Mysqldb()
```

(4)打开应用 djangoProject 中的配置文件 settings.py,进行数据库的相关配置,用于连接数据库,代码如下:

```python
# 资源包\Code\chapter5\5.7\3\djangoProject\djangoProject\settings.py
DATABASES = {
    'default': {
        # 数据库引擎
        'ENGINE': 'django.db.backends.mysql',
        # 数据库名称
        'NAME': 'django_db',
        # 连接 MySQL 的用户名
        'USER': 'root',
        # 连接 MySQL 的密码
        'PASSWORD': '12345678',
        # MySQL 的主机地址
        'HOST': '127.0.0.1',
        # MySQL 的端口号
        'PORT': '3306',
    }
}
```

(5)打开应用 djangoProject 中的配置文件 settings.py,将应用 book 和 press 添加到变

量 INSTALLED_APPS 中,用于完成应用注册,代码如下:

```python
# 资源包\Code\chapter5\5.7\3\djangoProject\djangoProject\settings.py
INSTALLED_APPS = [
    'django.contrib.admin',
    'django.contrib.auth',
    'django.contrib.contenttypes',
    'django.contrib.sessions',
    'django.contrib.messages',
    'django.contrib.staticfiles',
    'book',
    'press',
]
```

(6) 打开应用 book 中的文件 models.py,代码如下:

```python
# 资源包\Code\chapter5\5.7\3\djangoProject\book\models.py
from django.db import models
# 父表
class Category(models.Model):
    cate = models.CharField(max_length = 100)
# 子表
class Book(models.Model):
    name = models.CharField(max_length = 100)
    author = models.CharField(max_length = 20, null = False)
    category = models.ForeignKey("Category", on_delete = models.CASCADE)
    # category = models.ForeignKey("Category", on_delete = models.PROTECT)
    # category = models.ForeignKey("Category", null = True, on_delete = models.SET_NULL)
    # category = models.ForeignKey("Category", null = True, on_delete = models.SET_DEFAULT, default = Category.objects.get(cate = 'Python 编程'))
    # category = models.ForeignKey("Category", null = True, on_delete = models.SET(Category.objects.get(cate = 'Django 编程')))
    press = models.ForeignKey('press.Press', on_delete = models.CASCADE, null = True)
```

(7) 打开应用 press 中的文件 models.py,代码如下:

```python
# 资源包\Code\chapter5\5.7\3\djangoProject\press\models.py
from django.db import models
class Press(models.Model):
    press_name = models.CharField(max_length = 100, null = True)
```

(8) 打开应用 book 中的文件 views.py,代码如下:

```python
# 资源包\Code\chapter5\5.7\3\djangoProject\book\views.py
from book.models import Book, Category
from press.models import Press
from django.http import HttpResponse
def add_data(request):
    category = Category(cate = '计算机编程')
    category.save()
    press = Press(press_name = '清华大学出版社')
    press.save()
    book = Book(name = '《Python 全栈开发——基础入门》', author = '夏正东')
    book.category = category
    book.press = press
```

```python
        book.save()
        return HttpResponse('新的图书已经添加成功!')
def del_category_data(request):
    category = Category.objects.get(cate = '计算机编程')
    category.delete()
    return HttpResponse('数据删除成功!')
#关联模型查询
def filter_data(request):
    categories = Category.objects.filter(book__name__contains = "Python")
    categories_lt = []
    for category in categories:
        categories_lt.append(category.cate)
    return HttpResponse(f'查询结果为{categories_lt}')
```

（9）打开应用djangoProject中的文件urls.py，代码如下：

```python
#资源包\Code\chapter5\5.7\3\djangoProject\djangoProject\urls.py
from django.urls import path
from book import views
urlpatterns = [
    path('add_data/', views.add_data),
    path('del_category_data/', views.del_category_data),
    path('filter_data/', views.filter_data),
]
```

（10）打开命令提示符窗口，登入MySQL。

（11）在MySQL命令行窗口中输入SQL语句"create database django_db;"，创建数据库django_db，其结果如图5-116所示。

（12）打开命令提示符窗口，进入项目所在的根目录中，激活虚拟环境，并输入命令python manage.py makemigrations，生成迁移脚本文件，如图5-117所示。

图5-116　创建数据库 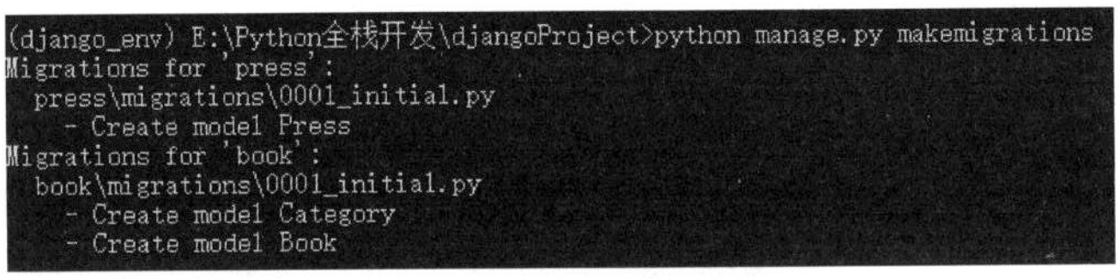　　　　　图5-117　生成迁移脚本文件

（13）再次输入命令python manage.py migrate，将迁移脚本文件映射到数据库中，如图5-118所示。

（14）在MySQL命令行窗口中输入SQL语句"desc book_book;"，查询该数据表的表结构，其结果如图5-119所示。

（15）再次输入SQL语句"desc book_category;"，查询该数据表的表结构，其结果如图5-120所示。

（16）再次输入SQL语句"desc press_press;"，查询该数据表的表结构，其结果如图5-121所示。

（17）运行上述项目，打开浏览器并在网址栏输入http://127.0.0.1:8000/add_data/，其显示内容如图5-122所示。

图 5-118　将迁移脚本文件映射到数据库中

图 5-119　数据表 book_book 的表结构

图 5-120　数据表 book_category 的表结构

图 5-121　数据表 press_press 的表结构

（18）在浏览器中的网址栏输入 http://127.0.0.1:8000/filter_data/，其显示内容如图 5-123 所示。

图 5-122　浏览器中的显示内容(1)　　　　图 5-123　浏览器中的显示内容(2)

（19）在 MySQL 命令行窗口中输入 SQL 语句"select * from book_book;"，其结果如图 5-124 所示。

图 5-124　数据表 book_book 中的数据

（20）输入 SQL 语句"select * from book_category;"，其结果如图 5-125 所示。
（21）输入 SQL 语句"select * from press_press;"，其结果如图 5-126 所示。

图 5-125　数据表 book_category 中的数据　　图 5-126　数据表 press_press 中的数据

（22）此时，在浏览器中的网址栏输入 http://127.0.0.1:8000/del_category_data/ 时，不同的外键约束会显示不同的运行的结果。

一是当外键约束为 models.CASCADE 时，在浏览器中的网址栏输入 http://127.0.0.1:8000/del_category_data/，其显示内容如图 5-127 所示。

图 5-127　浏览器中的显示内容

此时，在 MySQL 命令行窗口中输入 SQL 语句"select * from book_book;"，其结果如图 5-128 所示。

再次输入 SQL 语句"select * from book_category;"，其结果如图 5-129 所示。

图 5-128　数据表 book_book 中的数据　　图 5-129　数据表 book_category 中的数据

二是当外键约束为 models.PROTECT 时，在浏览器中的网址栏输入 http://127.0.0.1:8000/del_category_data/，其显示内容如图 5-130 所示。

图 5-130　浏览器中的显示内容

此时，在 MySQL 命令行窗口中输入 SQL 语句"select * from book_book;"，如图 5-131 所示。

图 5-131　数据表 book_book 中的数据

再次输入 SQL 语句"select * from book_category;"，如图 5-132 所示。

三是当外键约束为 models.SET_NULL 时，在浏览器中的网址栏输入 http://127.0.0.1:8000/del_category_data/，其显示内容如图 5-133 所示。

图 5-132　数据表 book_category 中的数据

图 5-133　浏览器中的显示内容

此时，在 MySQL 命令行窗口中输入 SQL 语句"select * from book_book;"，其结果如图 5-134 所示。

图 5-134　数据表 book_book 中的数据

再次输入 SQL 语句"select * from book_category;"，其结果如图 5-135 所示。

四是当外键约束为 models.SET_DEFAULT 时，在 MySQL 命令行窗口中输入 SQL 语句"insert into book_category (cate) values ('Python 编程');"，添加新的数据，然后再次输入 SQL 语句"select * from book_category;"，查看数据表中的数据，其结果如图 5-136 所示。

图 5-135　数据表 book_category 中的数据

图 5-136　数据表 book_category 中的数据

需要注意的是，必须先在数据表 book_category 中添加数据，然后在浏览器中的网址栏输入 http://127.0.0.1:8000/add_data/，添加其他数据；最后，在浏览器中的网址栏输入 http://127.0.0.1:8000/del_category_data/，其显示内容如图 5-137 所示。

此时，在 MySQL 命令行窗口中输入 SQL 语句"select * from book_book;"，其结果如图 5-138 所示。

再次输入 SQL 语句"select * from book_category;"，其结果如图 5-139 所示。

图 5-137　浏览器中的显示内容

图 5-138　数据表 book_book 中的数据

五是当外键约束为 models.SET 时,在 MySQL 命令行窗口中输入 SQL 语句"insert into book_category (cate) values ('Django编程');",添加新的数据,然后再次输入 SQL 语句"select * from book_category;",查看数据表中的数据,如图 5-140 所示。

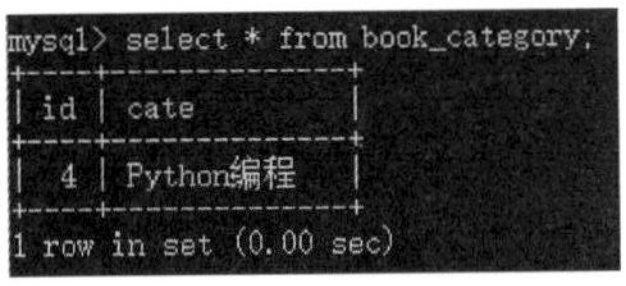

图 5-139　数据表 book_category 中的数据 　　图 5-140　数据表 book_category 中的数据

需要注意的是,必须先在数据表 book_category 添加数据,然后在浏览器中的网址栏输入 http://127.0.0.1:8000/add_data/,添加其他数据;最后,在浏览器中的网址栏输入 http://127.0.0.1:8000/del_category_data/,其显示内容如图 5-141 所示。

图 5-141　浏览器中的显示内容

此时,在 MySQL 命令行窗口中输入 SQL 语句"select * from book_book;",其结果如图 5-142 所示。

图 5-142　数据表 book_book 中的数据

再次输入 SQL 语句"select * from book_category;",其结果如图 5-143 所示。

图 5-143　数据表 book_category 中的数据

5.7.7 多表间关系

1. 一对多关系

一对多关系表示一个数据模型可以对应多个其他数据模型，而其他数据模型只能对应一个数据模型，例如，有两个数据模型 User 和 Book，即一位作者可以编写多本书籍，而一本书籍只能有一位作者。可以通过数据模型中的 ForeignKey 字段实现。

下面通过一个示例演示一下如何实现一对多关系。

（1）创建名为 djangoProject 的 Django 项目。

（2）打开命令提示符窗口，进入项目所在的根目录中，激活虚拟环境，并输入命令 python manage. py startapp book，创建名为 book 的应用，如图 5-144 所示。

```
E:\Python全栈开发\djangoProject>workon django_env
(django_env) E:\Python全栈开发\djangoProject>python manage.py startapp book
```

图 5-144 创建应用

（3）打开应用 djangoProject 中的文件__init__. py，代码如下：

```
#资源包\Code\chapter5\5.7\4\djangoProject\djangoProject\__init__.py
import pymysql
pymysql.install_as_Mysqldb()
```

（4）打开应用 djangoProject 中的配置文件 settings. py，进行数据库的相关配置，用于连接数据库，代码如下：

```
#资源包\Code\chapter5\5.7\4\djangoProject\djangoProject\settings.py
DATABASES = {
    'default': {
        #数据库引擎
        'ENGINE': 'django.db.backends.mysql',
        #数据库名称
        'NAME': 'django_db',
        #连接 MySQL 的用户名
        'USER': 'root',
        #连接 MySQL 的密码
        'PASSWORD': '12345678',
        #MySQL 的主机地址
        'HOST': '127.0.0.1',
        #MySQL 的端口号
        'PORT': '3306',
    }
}
```

（5）打开应用 djangoProject 中的配置文件 settings. py，将应用 book 添加到变量 INSTALLED_APPS 中，用于完成应用注册，代码如下：

```
#资源包\Code\chapter5\5.7\4\djangoProject\djangoProject\settings.py
INSTALLED_APPS = [
    'django.contrib.admin',
    'django.contrib.auth',
    'django.contrib.contenttypes',
```

```
    'django.contrib.sessions',
    'django.contrib.messages',
    'django.contrib.staticfiles',
    'book',
]
```

（6）打开应用 book 中的文件 models.py，代码如下：

```
#资源包\Code\chapter5\5.7\4\djangoProject\book\models.py
from django.db import models
class Author(models.Model):
    name = models.CharField(max_length = 100)
class Book(models.Model):
    name = models.CharField(max_length = 100)
    press = models.CharField(max_length = 20, null = False)
    #反向引用的名称为 books
    author = models.ForeignKey("Author", on_delete = models.CASCADE, related_name = 'books')
```

（7）打开应用 book 中的文件 views.py，代码如下：

```
#资源包\Code\chapter5\5.7\4\djangoProject\book\views.py
from book.models import Book, Author
from django.http import HttpResponse
def one_to_many_set(request):
    author = Author(name = '夏正东')
    author.save()
    book1 = Book(name = '《Python全栈开发——基础入门》', press = '清华大学出版社')
    book1.author = author
    book1.save()
    book2 = Book(name = '《Python全栈开发——高阶编程》', press = '清华大学出版社')
    book2.author = author
    book2.save()
    return HttpResponse('新的图书已经添加成功！')
def one_to_many_get(request):
    author = Author.objects.first()
    #获取当前作者所编写的所有书籍，即反向引用
    books = author.books.all()
    book_lt = []
    for book in books:
        book_lt.append(book.name)
    return HttpResponse(book_lt)
```

（8）打开应用 djangoProject 中的文件 urls.py，代码如下：

```
#资源包\Code\chapter5\5.7\4\djangoProject\djangoProject\urls.py
from django.urls import path
from book import views
urlpatterns = [
    path('one_to_many_set/', views.one_to_many_set),
    path('one_to_many_get/', views.one_to_many_get),
]
```

（9）打开命令提示符窗口，登入 MySQL。

（10）在 MySQL 命令行窗口中输入 SQL 语句"create database django_db;"，创建数据

库 django_db，其结果如图 5-145 所示。

（11）打开命令提示符窗口，进入项目所在的根目录中，激活虚拟环境，并输入命令 python manage.py makemigrations，生成迁移脚本文件，如图 5-146 所示。

图 5-145　创建数据库　　　　　　　　图 5-146　生成迁移脚本文件

（12）再次输入命令 python manage.py migrate，将迁移脚本文件映射到数据库中，如图 5-147 所示。

（13）运行上述项目，打开浏览器并在网址栏输入 http://127.0.0.1:8000/one_to_many_set/，其显示内容如图 5-148 所示。

图 5-147　迁移脚本文件映射到数据库中　　　　图 5-148　浏览器中的显示内容

（14）在 MySQL 命令行窗口中输入 SQL 语句"select * from book_book;"，其结果如图 5-149 所示。

图 5-149　数据表 book_book 中的数据

（15）再次输入 SQL 语句"select * from book_author;"，其结果如图 5-150 所示。

（16）在浏览器中的网址栏输入 http://127.0.0.1:8000/one_to_many_get/，其显示内容如图 5-151 所示。

图 5-150　数据表 book_author 中的数据　　　　图 5-151　浏览器中的显示内容

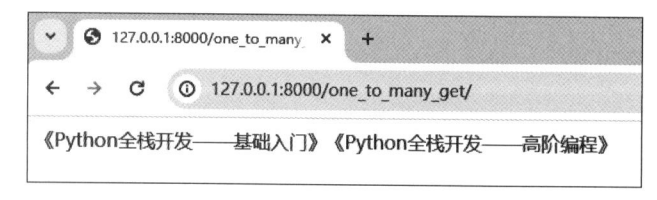

2. 一对一关系

一对一关系表示两个数据模型之间存在一个唯一的对应关系，例如，有两个数据模型

User 和 UserExtend，即一个用户只能有一个该用户的扩展，而扩展也只能对应一个用户。可以通过数据模型中的 OneToOneField 字段实现。

下面通过一个示例，演示一下如何实现一对一关系。

(1) 创建名为 djangoProject 的 Django 项目。

(2) 打开命令提示符窗口，进入项目所在的根目录中，激活虚拟环境，并输入命令 python manage.py startapp user，创建名为 user 的应用，如图 5-152 所示。

```
E:\Python全栈开发\djangoProject>workon django_env
(django_env) E:\Python全栈开发\djangoProject>python manage.py startapp user
```

图 5-152　创建应用

(3) 打开应用 djangoProject 中的文件 __init__.py，代码如下：

```
# 资源包\Code\chapter5\5.7\5\djangoProject\djangoProject\__init__.py
import pymysql
pymysql.install_as_Mysqldb()
```

(4) 打开应用 djangoProject 中的配置文件 settings.py，进行数据库的相关配置，用于连接数据库，代码如下：

```
# 资源包\Code\chapter5\5.7\5\djangoProject\djangoProject\settings.py
DATABASES = {
    'default': {
        # 数据库引擎
        'ENGINE': 'django.db.backends.mysql',
        # 数据库名称
        'NAME': 'django_db',
        # 连接MySQL的用户名
        'USER': 'root',
        # 连接MySQL的密码
        'PASSWORD': '12345678',
        # MySQL的主机地址
        'HOST': '127.0.0.1',
        # MySQL的端口号
        'PORT': '3306',
    }
}
```

(5) 再次打开应用 djangoProject 中的配置文件 settings.py，将应用 user 添加到变量 INSTALLED_APPS 中，用于完成应用注册，代码如下：

```
# 资源包\Code\chapter5\5.7\5\djangoProject\djangoProject\settings.py
INSTALLED_APPS = [
    'django.contrib.admin',
    'django.contrib.auth',
    'django.contrib.contenttypes',
    'django.contrib.sessions',
    'django.contrib.messages',
    'django.contrib.staticfiles',
    'user',
]
```

(6) 打开应用 user 中的文件 models.py，代码如下：

```
# 资源包\Code\chapter5\5.7\5\djangoProject\user\models.py
from django.db import models
class User(models.Model):
    username = models.CharField(max_length=100)
class UserExtend(models.Model):
    job = models.CharField(max_length=100)
    user = models.OneToOneField("User", on_delete=models.CASCADE, related_name='userextend')
```

(7) 打开应用 user 中的文件 views.py,代码如下:

```
# 资源包\Code\chapter5\5.7\5\djangoProject\user\views.py
from user.models import User, UserExtend
from django.http import HttpResponse
def one_to_one_set(request):
    user = User(username='夏正东')
    user.save()
    userextend = UserExtend(job='老师')
    userextend.user = user
    userextend.save()
    return HttpResponse('新的用户信息已经添加成功!')
def one_to_one_get(request):
    user = User.objects.first()
    # 获取当前用户所对应的扩展信息,即反向引用
    userextend = user.userextend
    return HttpResponse(f'当前用户的职业是: {userextend.job}')
```

(8) 打开应用 djangoProject 中的文件 urls.py,代码如下:

```
# 资源包\Code\chapter5\5.7\5\djangoProject\djangoProject\urls.py
from django.urls import path
from user import views
urlpatterns = [
    path('one_to_one_set/', views.one_to_one_set),
    path('one_to_one_get/', views.one_to_one_get),
]
```

(9) 打开命令提示符窗口,登入 MySQL。

(10) 在 MySQL 命令行窗口中输入 SQL 语句"create database django_db;",创建数据库 django_db,其结果如图 5-153 所示。

(11) 打开命令提示符窗口,进入项目所在的根目录中,激活虚拟环境,并输入命令 python manage.py makemigrations,生成迁移脚本文件,如图 5-154 所示。

图 5-153　创建数据库　　　　　　图 5-154　生成迁移脚本文件

(12) 再次输入命令 python manage.py migrate,将迁移脚本文件映射到数据库中,如图 5-155 所示。

(13) 运行上述项目,打开浏览器并在网址栏输入 http://127.0.0.1:8000/one_to_one_

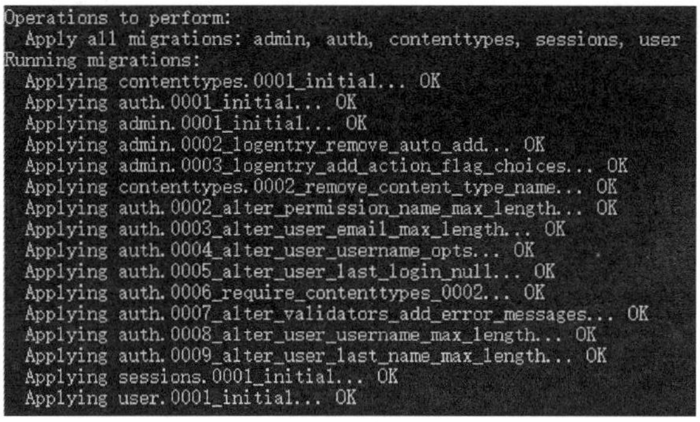

图 5-155　迁移脚本文件映射到数据库中

set/，其显示内容如图 5-156 所示。

图 5-156　浏览器中的显示内容

（14）在 MySQL 命令行窗口中输入 SQL 语句"select * from user_user;"，其结果如图 5-157 所示。

（15）再次输入 SQL 语句"select * from user_userextend;"，其结果如图 5-158 所示。

图 5-157　数据表 user_user 中的数据　　图 5-158　数据表 user_userextend 中的数据

（16）在浏览器中的网址栏输入 http://127.0.0.1:8000/one_to_one_get/，其显示内容如图 5-159 所示。

图 5-159　浏览器中的显示内容

3．多对多关系

多对多关系表示两个数据模型之间存在多个对应关系，例如，有两个数据模型 Article 和 Tag，即一篇文章可以对应多个标签，并且一个标签也可以对应多篇文章。可以通过数据模型中的 ManyToManyField 字段实现。

下面通过一个示例演示一下如何实现多对多关系。

（1）创建名为 djangoProject 的 Django 项目。

（2）打开命令提示符窗口，进入项目所在的根目录中，激活虚拟环境，并输入命令 python manage.py startapp article，创建名为 article 的应用，如图 5-160 所示。

```
(django_env) E:\Python全栈开发\djangoProject>python manage.py startapp article
```

图 5-160　创建应用

(3) 打开应用 djangoProject 中的文件 __init__.py，代码如下：

```python
# 资源包\Code\chapter5\5.7\6\djangoProject\djangoProject\__init__.py
import pymysql
pymysql.install_as_Mysqldb()
```

(4) 打开应用 djangoProject 中的配置文件 settings.py，进行数据库的相关配置，用于连接数据库，代码如下：

```python
# 资源包\Code\chapter5\5.7\6\djangoProject\djangoProject\settings.py
DATABASES = {
    'default': {
        # 数据库引擎
        'ENGINE': 'django.db.backends.mysql',
        # 数据库名称
        'NAME': 'django_db',
        # 连接 MySQL 的用户名
        'USER': 'root',
        # 连接 MySQL 的密码
        'PASSWORD': '12345678',
        # MySQL 的主机地址
        'HOST': '127.0.0.1',
        # MySQL 的端口号
        'PORT': '3306',
    }
}
```

(5) 打开应用 djangoProject 中的配置文件 settings.py，将应用 article 添加到变量 INSTALLED_APPS 中，用于完成应用注册，代码如下：

```python
# 资源包\Code\chapter5\5.7\6\djangoProject\djangoProject\settings.py
INSTALLED_APPS = [
    'django.contrib.admin',
    'django.contrib.auth',
    'django.contrib.contenttypes',
    'django.contrib.sessions',
    'django.contrib.messages',
    'django.contrib.staticfiles',
    'article',
]
```

(6) 打开应用 article 中的文件 models.py，代码如下：

```python
# 资源包\Code\chapter5\5.7\6\djangoProject\article\models.py
from django.db import models
class Article(models.Model):
    title = models.CharField(max_length=100)
    username = models.CharField(max_length=100)
class Tag(models.Model):
    name = models.CharField(max_length=100)
    articles = models.ManyToManyField("Article", related_name='tags')
```

(7) 打开应用 article 中的文件 views.py，代码如下：

```python
# 资源包\Code\chapter5\5.7\6\djangoProject\article\views.py
from article.models import Article, Tag
from django.http import HttpResponse
def many_to_many_set(request):
    tag1 = Tag(name = '计算机文章')
    tag1.save()
    tag2 = Tag(name = '热门文章')
    tag2.save()
    article1 = Article(title = 'Django开发的流程', username = '夏正东')
    article1.save()
    article2 = Article(title = '如何养成缜密的逻辑思维', username = '夏正东')
    article2.save()
    # 通过 add()方法添加数据
    article1.tags.add(tag1)
    article1.tags.add(tag2)
    article2.tags.add(tag2)
    return HttpResponse('新的文章已经添加成功！')
def many_to_many_get(request):
    article = Article.objects.first()
    # 获取当前用户所对应的扩展信息，即反向引用
    tags = article.tags.all()
    tag_lt = []
    for tag in tags:
        tag_lt.append(tag.name)
    return HttpResponse(tag_lt)
```

(8) 打开应用 djangoProject 中的文件 urls.py，代码如下：

```python
# 资源包\Code\chapter5\5.7\6\djangoProject\djangoProject\urls.py
from django.urls import path
from article import views
urlpatterns = [
    path('many_to_many_set/', views.many_to_many_set),
    path('many_to_many_get/', views.many_to_many_get),
]
```

(9) 打开命令提示符窗口，登入 MySQL。

(10) 在 MySQL 命令行窗口中输入 SQL 语句"create database django_db;"，创建数据库 django_db，其结果如图 5-161 所示。

(11) 打开命令提示符窗口，进入项目所在的根目录中，激活虚拟环境，并输入命令 python manage.py makemigrations，生成迁移脚本文件，如图 5-162 所示。

图 5-161　创建数据库 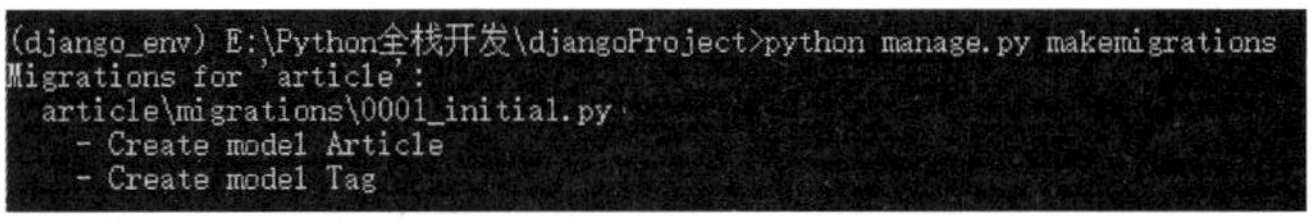　　　　图 5-162　生成迁移脚本文件

(12) 再次输入命令 python manage.py migrate，将迁移脚本文件映射到数据库中，如图 5-163 所示。

(13) 运行上述项目，打开浏览器并在网址栏输入 http://127.0.0.1:8000/many_to_many_set/，其显示内容如图 5-164 所示。

图 5-163　迁移脚本文件映射到数据库中

（14）在 MySQL 命令行窗口中输入 SQL 语句"select * from article_article;"，其结果如图 5-165 所示。

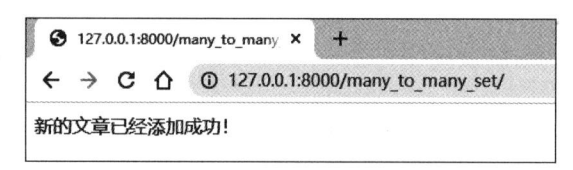

图 5-164　浏览器中的显示内容

图 5-165　表 article_article 中的数据

（15）再次输入 SQL 语句"select * from article_tag;"，其结果如图 5-166 所示。

（16）此时，Django 会在数据库中自动生成关联表 article_article_tag，输入 SQL 语句"select * from article_article_tag;"，其结果如图 5-167 所示。

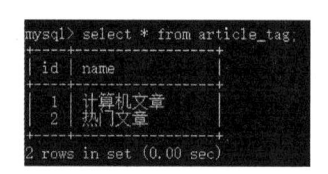

图 5-166　表 article_tag 中的数据

图 5-167　表 article_article_tag 中的数据

（17）在浏览器中的网址栏输入 http://127.0.0.1:8000/many_to_many_get/，其显示内容如图 5-168 所示。

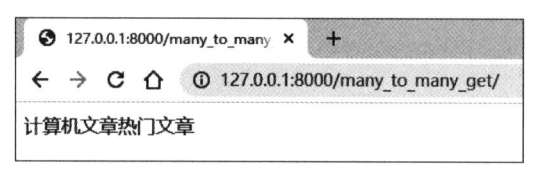

图 5-168　浏览器中的显示内容

5.8　表单验证

5.8.1　HTML 表单验证

在 Django 中表单验证可以分为 4 步：

(1) 导入 django.forms 模块和 django.core.validators 模块中的相关类,其中,django.forms 模块包含支持 HTML 的标准字段类(如表 5-11 所示);django.core.validators 模块包含对 HTML 表单进行验证的验证器类,包括内置验证器类和自定义验证器类。

表 5-11 HTML 标准字段类

字 段 类	参 数	描 述
CharField	label,表示字段别名;	字符串字段
FloatField	validators,表示验证规则组成的	浮点数字段
IntegerField	列表;	整型字段
EmailField	help_text,表示字段帮助信息;	电子邮件地址字段
URLField	required,表示是否为必填字段;	URL 字段
FileField	widget,表示 HTML 插件;	文件上传字段
ImageField	error_messages,表示自定义错误的	图片字段
BooleanField	字典;	复选框字段
ChoiceField	disabled,表示是否为禁用字段;	下拉列表字段
MultipleChoiceField	label_suffix;表示所有字段的标签	多选下拉列表字段
DateField	后缀,在默认情况下为冒号;	日期字段
DateTimeField	initial,表示字段的初始值	日期时间字段

(2) 创建表单类:表单类主要用于定义 HTML 表单中的字段,该类中的属性对应 HTML 表单中的每个字段。需要注意的是,表单类必须继承自 django.forms 中的 Form 类,并且其属性必须与表单字段中属性 name 的值一致。

(3) 处理验证数据:在视图中,通过表单对象的 is_valid() 方法处理数据验证之后的结果。

(4) 指定验证器:在验证表单中的字段时,还可以通过给字段类中的参数 validators 传递指定验证器,从而进一步对数据进行过滤,其常用的内置验证器类如表 5-12 所示。

表 5-12 常用的内置验证器类

内置验证器类	描 述
MaxValueValidator	用于确保数据的值不超过指定的最大值
MinValueValidator	用于确保数据的值不低于指定的最小值
MinLengthValidator	用于验证字段输入长度是否达到最小长度
MaxLengthValidator	用于确保字段的长度不会超过指定的最大长度
EmailValidator	用于验证电子邮件地址
URLValidator	用于验证输入是否为有效的 URL
RegexValidator	用于确保字段的输入匹配一个正则表达式

其中,自定义验证器类需要在表单类中创建一种方法,如果验证单一字段,则该方法的命名规则需为"clean_字段名",如果验证多字段,则该方法的命名规则需为 clean。

下面通过一个示例,演示一下如何进行 HTML 表单验证。

(1) 创建名为 djangoProject 的 Django 项目。

(2) 打开命令提示符窗口,进入项目所在的根目录中,激活虚拟环境,并输入命令 python manage.py startapp book,创建名为 book 的应用,如图 5-169 所示。

(3) 在应用 book 中创建表单验证文件 forms.py,代码如下:

```
E:\Python全栈开发\djangoProject>workon django_env
(django_env) E:\Python全栈开发\djangoProject>python manage.py startapp book
```

图 5-169　创建应用

```python
# 资源包\Code\chapter5\5.8\1\djangoProject\book\forms.py
from django import forms
from django.core import validators
class RegForm(forms.Form):
    book_name = forms.CharField(max_length=8, min_length=2)
    book_price = forms.FloatField(max_value=200, min_value=50)
    author_tel = forms.CharField(validators=[validators.RegexValidator(r'1[3456789]\d{9}')])
    author_email = forms.EmailField()
    web_site = forms.urlField()
    pwd1 = forms.CharField(max_length=10, min_length=6)
    pwd2 = forms.CharField(max_length=10, min_length=6)
    # 单一字段验证(自定义)
    def clean_author_email(self):
        # cleaned_data 含通过验证的表单数据所组成的字典
        author_email = self.cleaned_data.get('author_email')
        if author_email == 'oldxia1@qq.com':
            raise forms.ValidationError(message='此邮箱不被允许')
        return author_email
    # 多字段验证(自定义)
    def clean(self):
        result = super().clean()
        pwd1 = self.cleaned_data.get('pwd1')
        pwd2 = self.cleaned_data.get('pwd2')
        if pwd1 != pwd2:
            raise forms.ValidationError(message='两次密码不一致')
        return result
```

（4）打开应用 book 中的文件 views.py，代码如下：

```python
# 资源包\Code\chapter5\5.8\1\djangoProject\book\views.py
from django.http import HttpResponse
from django.shortcuts import render
from django.views.generic import View
from book.forms import RegForm
class RegView(View):
    def get(self, request):
        return render(request, 'index.html')
    def post(self, request):
        form = RegForm(request.POST)
        # 处理验证数据
        if form.is_valid():
            return HttpResponse('数据已经提交!')
        else:
            return HttpResponse('提交的数据有误,请重新提交!')
```

（5）在项目根目录下的文件夹 templates 中创建模板文件 index.html，代码如下：

```
# 资源包\Code\chapter5\5.8\1\djangoProject\templates\index.html
<html lang="en">
    <head>
```

```html
        <meta charset="UTF-8">
        <title>Title</title>
    </head>
    <body>
        <form action="" method="post">
            {% csrf_token %}
            <p>书籍名称：<input type="text" name="book_name"></p>
            <p>书籍价格：<input type="text" name="book_price"></p>
            <p>作者手机号码：<input type="text" name="author_tel"></p>
            <p>作者Email：<input type="text" name="author_email"></p>
            <p>作者个人博客：<input type="text" name="web_site"></p>
            <p>密码1：<input type="password" name="pwd1"></p>
            <p>密码2：<input type="password" name="pwd2"></p>
            <input type="submit" value="提交">
        </form>
    </body>
</html>
```

（6）打开应用djangoProject中的文件urls.py，代码如下：

```
# 资源包\Code\chapter5\5.8\1\djangoProject\djangoProject\urls.py
from django.urls import path
from book import views
urlpatterns = [
    path('', views.RegView.as_view()),
]
```

（7）运行上述项目，打开浏览器并在网址栏输入http://127.0.0.1:8000/，其显示内容如图5-170所示。

（8）此时，在表单中填入相关数据，其显示内容如图5-171所示。

图5-170　浏览器中的显示内容(1)　　图5-171　浏览器中的显示内容(2)

由于填入的"作者手机号码"格式错误，所以表单数据无法成功提交，其显示内容如图5-172所示。

图 5-172　浏览器中的显示内容(3)

5.8.2　上传文件验证

上传文件验证需要注意两点：一是表单类中对应属性的字段类型需为 FileField；二是通过验证器类中的参数 allowed_extensions，可以对文件类型进行限制。

下面通过一个示例演示一下如何进行上传文件验证。

（1）创建名为 djangoProject 的 Django 项目。

（2）打开命令提示符窗口，进入项目所在的根目录中，激活虚拟环境，并输入命令 python manage.py startapp book，创建名为 book 的应用，如图 5-173 所示。

图 5-173　创建应用

（3）在应用 book 中创建表单验证文件 forms.py，代码如下：

```python
# 资源包\Code\chapter5\5.8\2\djangoProject\book\forms.py
from django import forms
from django.core import validators
class UploadFileForm(forms.Form):
    myfile = forms.FileField(validators=[validators.FileExtensionValidator(allowed_extensions=['pdf','docx','doc'])])
```

（4）打开应用 book 中的文件 views.py，代码如下：

```python
# 资源包\Code\chapter5\5.8\2\djangoProject\book\views.py
from django.views.generic import View
from django.shortcuts import render
from django.http import HttpResponse
from book.forms import UploadFileForm
class IndexView(View):
    def get(self, request):
        return render(request, 'index.html')
    def post(self, request):
        form = UploadFileForm(request.POST, request.FILES)
        if form.is_valid():
            myfile = request.FILES.get('myfile')
            # 此时,myfile 类型为 InMemoryUploadedFile,可以通过 read()方法读取文件内容,并
            # 返回二进制数据
            with open('file.txt', 'wb') as fb:
                fb.write(myfile.read())
            return HttpResponse('电子书上传成功!')
        else:
            return HttpResponse('电子书上传失败!')
```

（5）在项目根目录下的文件夹 templates 中创建模板文件 index.html，代码如下：

```
# 资源包\Code\chapter5\5.8\2\djangoProject\templates\index.html
<!DOCTYPE html>
<html lang="en">
    <head>
        <meta charset="UTF-8">
        <title>Title</title>
    </head>
    <body>
        <form action="" method="post" enctype="multipart/form-data">
            {% csrf_token %}
            <p>文件:<input type="file" name="myfile" id=""></p>
            <input type="submit" value="提交">
        </form>
    </body>
</html>
```

(6) 打开应用 djangoProject 中的文件 urls.py,代码如下:

```
# 资源包\Code\chapter5\5.8\2\djangoProject\djangoProject\urls.py
from django.urls import path
from book import views
urlpatterns = [
    path('', views.IndexView.as_view()),
]
```

(7) 运行上述项目,打开浏览器并在网址栏输入 http://127.0.0.1:8000/,其显示内容如图 5-174 所示。

(8) 此时,上传 PDF 文件并单击"提交"按钮,其显示内容如图 5-175 所示。

图 5-174　浏览器中的显示内容(1)　　　图 5-175　浏览器中的显示内容(2)

5.8.3　ModelForm 类

ModelForm 类是一种自动生成表单的工具,其是以数据模型为基础并在数据模型类上定义的表单。

在使用 ModelForm 类时,只需指定数据模型类作为表单数据的基础,就可以自动生成表单并进行验证。

下面通过一个示例,演示一下如何使用 ModelForm 类。

(1) 创建名为 djangoProject 的 Django 项目。

(2) 打开命令提示符窗口,进入项目所在的根目录中,激活虚拟环境,并输入命令 python manage.py startapp book,创建名为 book 的应用,如图 5-176 所示。

(3) 在项目根目录下创建文件夹 media,用于存放上传的文件。

```
E:\Python全栈开发\djangoProject>workon django_env
(django_env) E:\Python全栈开发\djangoProject>python manage.py startapp book
```

图 5-176　创建应用

(4) 在应用 book 中创建表单验证文件 forms.py，代码如下：

```python
# 资源包\Code\chapter5\5.8\3\djangoProject\book\forms.py
from django import forms
from book.models import Book
class AddBookForm(forms.ModelForm):
    def clean_author(self):
        author = self.cleaned_data.get('author')
        if author != '夏正东':
            print(author)
            raise forms.ValidationError('作者信息错误!')
        return author
    class Meta:
        model = Book
        # 需要验证的字段
        fields = "__all__"
        # 不需要验证的字段
        # Exclude = "author"
```

(5) 打开应用 book 中的文件 models.py，代码如下：

```python
# 资源包\Code\chapter5\5.8\3\djangoProject\book\models.py
from django.db import models
from django.core import validators
class Book(models.Model):
    name = models.CharField(max_length=100)
    author = models.CharField(max_length=20)
    price = models.FloatField(validators=[validators.MinValueValidator(limit_value=50)])
    # upload_to 定义上传文件存放的位置
    pdf = models.FileField(upload_to='%Y/%m/%d', validators=[validators.FileExtensionValidator(['pdf'])])
```

(6) 打开应用 book 中的文件 views.py，代码如下：

```python
# 资源包\Code\chapter5\5.8\3\djangoProject\book\views.py
from django.http import HttpResponse
from django.shortcuts import render
from book.forms import AddBookForm
from django.views.generic import View
class AddBookView(View):
    def get(self, request):
        return render(request, 'index.html')
    def post(selfm, request):
        form = AddBookForm(request.POST, request.FILES)
        myfile = request.FILES.get('pdf')
        print(myfile)
        if form.is_valid():
            form.save()
            return HttpResponse("书籍添加成功!")
        else:
            print(form.errors.get_json_data())
            return HttpResponse("书籍添加失败!")
```

(7) 在项目根目录下的文件夹 templates 中创建模板文件 index.html，代码如下：

```html
# 资源包\Code\chapter5\5.8\3\djangoProject\templates\index.html
<!DOCTYPE html>
<html lang="en">
    <head>
        <meta charset="UTF-8">
        <title>Title</title>
    </head>
    <body>
        <form action="" method="post" enctype="multipart/form-data">
            {% csrf_token %}
            <p>书籍名称:<input type="text" name="name"></p>
            <p>书籍作者:<input type="text" name="author"></p>
            <p>书籍价格:<input type="text" name="price"></p>
            <p>PDF书籍:<input type="file" name="pdf"></p>
            <p><input type="submit" value="添加书籍"></p>
        </form>
    </body>
</html>
```

(8) 打开应用 djangoProject 中的文件 urls.py，代码如下：

```python
# 资源包\Code\chapter5\5.8\3\djangoProject\djangoProject\urls.py
from Django.urls import path
from book import views
urlpatterns = [
    path('', views.AddBookView.as_view()),
]
```

(9) 打开应用 djangoProject 中的文件 __init__.py，代码如下：

```python
# 资源包\Code\chapter5\5.8\3\djangoProject\djangoProject\__init__.py
import pymysql
pymysql.install_as_Mysqldb()
```

(10) 打开应用 djangoProject 中的配置文件 settings.py，进行数据库的相关配置，用于连接数据库，代码如下：

```python
# 资源包\Code\chapter5\5.8\3\djangoProject\djangoProject\settings.py
DATABASES = {
    'default': {
        # 数据库引擎
        'ENGINE': 'django.db.backends.mysql',
        # 数据库名称
        'NAME': 'django_db',
        # 连接MySQL的用户名
        'USER': 'root',
        # 连接MySQL的密码
        'PASSWORD': '12345678',
        # MySQL的主机地址
        'HOST': '127.0.0.1',
        # MySQL的端口号
        'PORT': '3306',
    }
}
```

(11) 打开应用 djangoProject 中的配置文件 settings.py,将应用 book 添加到变量 INSTALLED_APPS 中,用于完成应用注册,代码如下:

```
#资源包\Code\chapter5\5.8\3\djangoProject\djangoProject\settings.py
INSTALLED_APPS = [
    'django.contrib.admin',
    'django.contrib.auth',
    'django.contrib.contenttypes',
    'django.contrib.sessions',
    'django.contrib.messages',
    'django.contrib.staticfiles',
    'book',
]
```

(12) 打开应用 djangoProject 中的配置文件 settings.py,设置上传文件的位置和访问上传文件的 URL,代码如下:

```
#资源包\Code\chapter5\5.8\3\djangoProject\djangoProject\settings.py
#上传文件的位置
MEDIA_ROOT = os.path.join(BASE_DIR, 'media')
#访问上传文件的 URL
MEDIA_URL = '/media/'
```

(13) 打开命令提示符窗口,进入项目所在的根目录中,激活虚拟环境,并输入命令 python manage.py makemigrations,生成迁移脚本文件,如图 5-177 所示。

图 5-177　生成迁移脚本文件

(14) 再次输入命令 python manage.py migrate,将迁移脚本文件映射到数据库中,如图 5-178 所示。

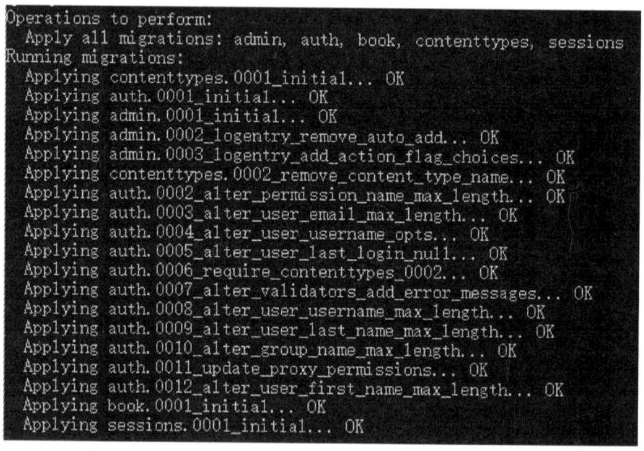

图 5-178　迁移脚本文件映射到数据库中

(15) 运行上述项目,打开浏览器并在网址栏输入 http://127.0.0.1:8000/,其显示内容如图 5-179 所示。

(16) 此时,在表单中填入相关数据,其显示内容如图 5-180 所示。

图 5-179　浏览器中的显示内容(1) 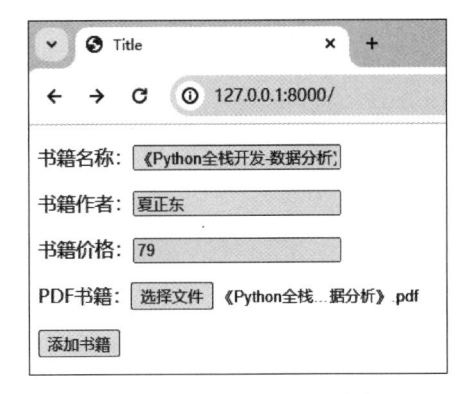　　　　图 5-180　浏览器中的显示内容(2)

单击"添加书籍"按钮后,其显示内容如图 5-181 所示。
上传文件在项目中的存储位置如图 5-182 所示。

图 5-181　浏览器中的显示内容(3) 　　　　图 5-182　上传文件的存储位置

5.9　Cookie 和 Session

5.9.1　设置、获取和删除 Cookie

1. 设置 Cookie

通过 HttpResponse 对象的 set_cookie()方法对 Cookie 进行设置,其语法格式如下:

```
set_cookie(key, value, max_age, expires, path, domain, secure, httponly)
```

其中,参数 key 表示 Cookie 的键,参数 value 表示 Cookie 的值,参数 max_age 表示 Cookie 被保存的时间,单位为秒,参数 expires 表示 Cookie 的具体过期时间,参数 path 用于限制 Cookie 的有效路径,参数 domain 用于设置 Cookie 可用的域名,参数 secure 用于设置 Cookie 是否仅通过 HTTPS 发送,参数 httponly 用于设置是否禁止 JavaScript 获取 Cookie。

2. 获取 Cookie

通过 HttpRequest 对象的属性 COOKIES 来获取指定 Cookie 的值。

3. 删除 Cookie

通过 HttpResponse 对象的 delete_cookie()方法对 Cookie 进行删除,其语法格式如下:

```
delete_cookie(key)
```

其中,参数 key 表示 Cookie 的键。

下面通过一个示例演示一下如何设置、获取和删除 Cookie。

(1) 创建名为 djangoProject 的 Django 项目。

(2) 打开命令提示符窗口,进入项目所在的根目录中,激活虚拟环境,并输入命令 python manage.py startapp book,创建名为 book 的应用,如图 5-183 所示。

图 5-183　创建应用

(3) 打开应用 book 中的文件 views.py,代码如下:

```python
# 资源包\Code\chapter5\5.9\1\djangoProject\book\views.py
from django.http import HttpResponse
def setcookie(request):
    response = HttpResponse('Cookie 已经设置成功!')
    response.set_cookie('web_site', 'http://www.oldxia.com')
    return response
def getcookie(request):
    cookies = request.COOKIES
    web_site = cookies.get('web_site')
    return HttpResponse(web_site)
def deletecookie(request):
    response = HttpResponse('Cookie 已经删除成功!')
    response.delete_cookie('web_site')
    return response
```

(4) 打开应用 djangoProject 中的文件 urls.py,代码如下:

```python
# 资源包\Code\chapter5\5.9\1\djangoProject\djangoProject\urls.py
from django.urls import path
from book import views
urlpatterns = [
    path('setcookie/', views.setcookie),
    path('getcookie/', views.getcookie),
    path('deletecookie/', views.deletecookie),
]
```

(5) 运行上述项目,打开浏览器并在网址栏输入 http://127.0.0.1:8000/setcookie/,即可设置 Cookie,其显示内容如图 5-184 所示。

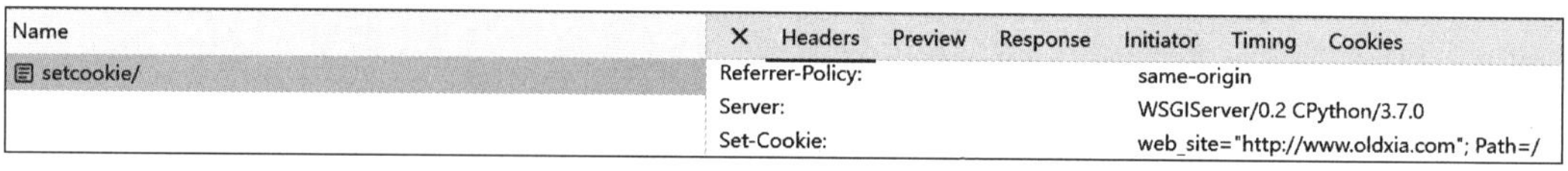

图 5-184　浏览器中的显示内容(1)

(6) 在浏览器中的网址栏输入 http://127.0.0.1:8000/getcookie/,即可获取 Cookie 的值,其显示内容如图 5-185 所示。

(7) 在浏览器中的网址栏输入 http://127.0.0.1:8000/deletecookie/,即可删除 Cookie,其显示内容如图 5-186 所示。

图 5-185　浏览器中的显示内容（2）

图 5-186　浏览器中的显示内容（3）

5.9.2　设置、获取和删除 Session

在默认情况下，由于 Session 存储在数据库的 django_session 数据表之中，所以在使用 Session 之前，需要在命令提示符中执行命令"python manage.py migrate"。

1. 设置 Session

通过 HttpRequest 对象的属性 session 对 Session 进行设置。

2. 获取 Session

通过 Session 对象的 get()方法来获取指定的 Session 值，其语法格式如下：

```
get(key)
```

其中，参数 key 表示 Session 的键。

3. 删除 Session

删除 Session 有两种方式。

一是通过 Session 对象的 pop()方法删除指定的 Session，其语法格式如下：

```
pop(key)
```

其中，参数 key 表示 Session 的键。

二是通过 Session 对象的 clear()方法删除全部的 Session，其语法格式如下：

```
clear()
```

下面通过一个示例演示一下如何设置、获取和删除 Session。

（1）创建名为 djangoProject 的 Django 项目。

（2）打开命令提示符窗口，进入项目所在的根目录中，激活虚拟环境，并输入命令 python manage.py startapp book，创建名为 book 的应用，如图 5-187 所示。

图 5-187　创建应用

（3）打开应用 book 中的文件 views.py，代码如下：

```
# 资源包\Code\chapter5\5.9\2\djangoProject\book\views.py
from django.http import HttpResponse
def setsession(request):
    request.session['web_site'] = 'http://www.oldxia.com'
    return HttpResponse('Session 已经设置成功！')
def getsession(request):
    web_site = request.session.get('web_site')
    return HttpResponse(web_site)
def deletesession(request):
    request.session.pop('web_site')
    return HttpResponse('Session 已经删除！')
```

（4）打开应用 djangoProject 中的文件 urls.py，代码如下：

```
# 资源包\Code\chapter5\5.9\2\djangoProject\djangoProject\urls.py
from django.urls import path
from book import views
urlpatterns = [
    path('setsession/', views.setsession),
    path('getsession/', views.getsession),
    path('deletesession/', views.deletesession),
]
```

（5）打开应用 djangoProject 中的文件 __init__.py，代码如下：

```
# 资源包\Code\chapter5\5.9\2\djangoProject\djangoProject\__init__.py
import pymysql
pymysql.install_as_Mysqldb()
```

（6）打开应用 djangoProject 中的配置文件 settings.py，进行数据库的相关配置，用于连接数据库，代码如下：

```
# 资源包\Code\chapter5\5.9\2\djangoProject\djangoProject\settings.py
DATABASES = {
    'default': {
        # 数据库引擎
        'ENGINE': 'django.db.backends.mysql',
        # 数据库名称
        'NAME': 'django_db',
        # 连接MySQL的用户名
        'USER': 'root',
        # 连接MySQL的密码
        'PASSWORD': '12345678',
        # MySQL的主机地址
        'HOST': '127.0.0.1',
        # MySQL的端口号
        'PORT': '3306',
    }
}
```

（7）打开命令提示符窗口，进入项目所在的根目录中，激活虚拟环境，并输入命令 python manage.py migrate，将迁移脚本文件映射到数据库中，如图 5-188 所示。

（8）运行上述项目，打开浏览器并在网址栏输入 http://127.0.0.1:8000/setsession/，

图 5-188　迁移脚本文件映射到数据库中

即可设置 Session，其显示内容如图 5-189 所示。

图 5-189　浏览器中的显示内容(1)

（9）在浏览器中的网址栏输入 http://127.0.0.1:8000/getsession/，即可获取 Session 的值，其显示内容如图 5-190 所示。

（10）在浏览器中的网址栏输入 http://127.0.0.1:8000/deletesession/，即可删除 Session。此时，在浏览器中的网址栏输入 http://127.0.0.1:8000/getsession/，其显示内容如图 5-191 所示。

图 5-190　浏览器中的显示内容(2)　　图 5-191　浏览器中的显示内容(3)

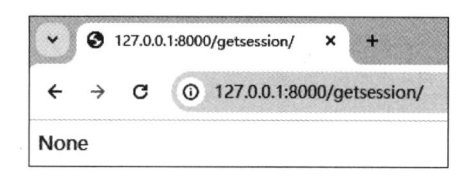

此外，可以在配置文件 settings.py 中添加变量 SESSION_ENGINE，用于对 Session 的存储机制进行修改。

（1）数据库存储（默认），代码如下：

```
SESSION_ENGINE = 'django.contrib.sessions.backends.db'
```

（2）文件存储，代码如下：

```
SESSION_ENGINE = 'django.contrib.sessions.backends.file'
```

（3）缓存存储，代码如下：

```
SESSION_ENGINE = 'django.contrib.sessions.backends.cache'
```

（4）缓存加数据库存储，代码如下：

```
SESSION_ENGINE = 'django.contrib.sessions.backends.cached_db'
```

(5) Cookie 存储，代码如下：

```
SESSION_ENGINE = 'django.contrib.sessions.backends.signed_cookies'
```

5.10 上下文处理器

上下文处理器是一种函数，用于在模板的上下文中添加额外的数据。

通过上下文处理器，开发人员可以方便地向模板中传递额外的数据，以便模板能够渲染所需的信息。

在 Django 中，上下文处理器分为内置上下文处理器和自定义上下文处理器。

1. 内置上下文处理器

在 Django 中，内置上下文处理器均在配置文件 settings.py 中的 TEMPLATES 配置的 OPTIONS 字典的 context_processors 列表中进行添加，其常用的内置上下文处理器如表 5-13 所示。

表 5-13 常用的内置上下文处理器

内置上下文处理器	描述
django.template.context_processors.debug	添加一个 debug 变量，当设置为 True 时，如果出现异常，则将在模板中显示详细的错误信息
django.template.context_processors.request	添加 HttpRequest 对象，用于在模板中获取一些与请求相关的信息，例如请求的方法、URL 或查询参数等
django.contrib.auth.context_processors.auth	添加一个 user 变量，用于在模板中添加有关当前用户的信息
django.contrib.messages.context_processors.messages	添加一个 messages 变量，用于在模板中添加错误、警告或任何需要展示给用户的消息

2. 自定义上下文处理器

自定义上下文处理器的步骤可以分为两步：一是在应用中创建一个名为 context_processors.py 的文件，用于自定义上下文处理器；二是在配置文件 settings.py 中注册该上下文处理器，即需要在配置文件 settings.py 中的 TEMPLATES 的 OPTIONS 字典的 context_processors 列表中添加该处理器。

下面通过一个示例演示一下如何使用上下文处理器。

（1）创建名为 djangoProject 的 Django 项目。

（2）打开命令提示符窗口，进入项目所在的根目录中，激活虚拟环境，并输入命令 python manage.py startapp book，创建名为 book 的应用，如图 5-192 所示。

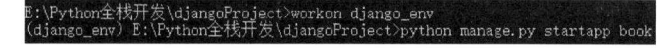

图 5-192 创建应用

（3）打开应用 book 中的文件 views.py，代码如下：

```
# 资源包\Code\chapter5\5.10\1\djangoProject\book\views.py
from django.shortcuts import render
```

```python
from django.contrib import messages
def index(request):
    messages.info(request, '这是内置上下文处理器')
    return render(request, 'index.html')
```

（4）在应用 book 中创建自定义上下文处理器文件 context_processors.py，代码如下：

```python
# 资源包\Code\chapter5\5.10\1\djangoProject\book\context_processors.py
import datetime
def get_daytime(request):
    now = datetime.datetime.now().strftime('%Y-%m-%d %H:%M:%S')
    my_hour = int(datetime.datetime.now().strftime('%H'))
    if my_hour >= 5 and my_hour <= 7:
        now_message = '早上好'
    elif my_hour > 7 and my_hour <= 11:
        now_message = '上午好'
    elif my_hour > 11 and my_hour < 18:
        now_message = '下午好'
    else:
        now_message = '晚上好'
    # 给模板传递所有变量
    return locals()
```

（5）在项目根目录下的文件夹 templates 中创建模板文件 index.html，代码如下：

```html
# 资源包\Code\chapter5\5.10\1\djangoProject\templates\index.html
<!DOCTYPE html>
<html lang="en">
    <head>
        <meta charset="UTF-8">
        <title>Title</title>
    </head>
    <body>
        <p>******* 内置上下文处理器 *******</p>
        <p>变量debug: {{ debug }}</p>
        <p>HttpRequest 对象: {{ request.method }}</p>
        <p>变量user: {{ user }}</p>
        <p>变量messages:
            {% for message in messages %}
                {{ message }}
            {% endfor %}
        </p>
        <p>******* 自定义上下文处理器 *******</p>
        <p>
            现在时间: {{ now }}<br>
            {{ now_message }}
        </p>
    </body>
</html>
```

（6）打开应用 djangoProject 中的配置文件 settings.py，对自定义上下文处理器进行注册，代码如下：

```python
# 资源包\Code\chapter5\5.10\1\djangoProject\djangoProject\settings.py
TEMPLATES = [
```

```
    {
        'BACKEND': 'django.template.backends.django.DjangoTemplates',
        'DIRS': [BASE_DIR / 'templates']
,
        'APP_DIRS': True,
        'OPTIONS': {
            'context_processors': [
                'django.template.context_processors.debug',
                'django.template.context_processors.request',
                'django.contrib.auth.context_processors.auth',
                'django.contrib.messages.context_processors.messages',
                'book.context_processors.get_daytime',
            ],
        },
    },
]
```

（7）打开应用 djangoProject 中的文件 urls.py，代码如下：

```
#资源包\Code\chapter5\5.10\1\djangoProject\djangoProject\urls.py
from django.urls import path
from book import views
urlpatterns = [
    path('', views.index),
]
```

（8）打开应用 djangoProject 中的文件 __init__.py，代码如下：

```
#资源包\Code\chapter5\5.10\1\djangoProject\djangoProject\__init__.py
import pymysql
pymysql.install_as_Mysqldb()
```

（9）打开应用 djangoProject 中的配置文件 settings.py，进行数据库的相关配置，用于连接数据库，代码如下：

```
#资源包\Code\chapter5\5.10\1\djangoProject\djangoProject\settings.py
DATABASES = {
    'default': {
        #数据库引擎
        'ENGINE': 'django.db.backends.mysql',
        #数据库名称
        'NAME': 'django_db',
        #连接 MySQL 的用户名
        'USER': 'root',
        #连接 MySQL 的密码
        'PASSWORD': '12345678',
        #MySQL 的主机地址
        'HOST': '127.0.0.1',
        #MySQL 的端口号
        'PORT': '3306',
    }
}
```

（10）打开命令提示符窗口，进入项目所在的根目录中，激活虚拟环境，并输入命令

python manage.py migrate,将迁移脚本文件映射到数据库中,如图 5-193 所示。

(11) 运行上述项目,打开浏览器并在网址栏输入 http://127.0.0.1:8000/,其显示内容如图 5-194 所示。

图 5-193　迁移脚本文件映射到数据库中

图 5-194　浏览器中的显示内容

5.11　中间件

中间件是一个轻量级、可重用的组件,用于处理 Django 请求和响应的过程,其提供了对请求和响应进行全局处理的机制,可以在请求达到视图之前进行预处理或在响应返回客户端之前进行后处理。

中间件是按顺序依次执行的,每个中间件都可以对请求和响应进行修改、补充或处理。

中间件通过配置文件 settings.py 中的变量 MIDDLEWARE 进行配置,包括注册和执行顺序。

在 Django 中,不仅内置了许多中间件,用于实现各自所对应的功能,同时还支持自定义中间件。

1. 内置中间件

内置中间件的作用主要包括认证和授权、请求和响应处理、异常处理和性能优化等,其常用的内置中间件如表 5-14 所示。

表 5-14　常用的内置中间件

内置中间件	描　　述
django.middleware.security.SecurityMiddleware	安全中间件负责处理与网站安全相关的任务
django.contrib.sessions.middleware.SessionMiddleware	会话中间件负责处理用户会话创建和检索用户数据
django.middleware.common.CommonMiddleware	通用中间件提供了一些常见而关键的 HTTP 请求处理功能
django.middleware.csrf.CsrfViewMiddleware	CSRF 中间件用于防止跨站请求伪造攻击
AuthenticationMiddleware	认证中间件负责处理用户身份认证相关的任务

续表

内置中间件	描述
MessageMiddleware	消息中间件用于在请求处理过程中存储和传递临时的一次性的用户消息
XFrameOptionsMiddleware	单击劫持中间件用于防止页面被嵌入其他网站中,从而提供一定的单击劫持保护

2. 自定义中间件

在 Django 中,可以通过函数或类进行中间件的自定义,并且必须将该自定义的中间件添加到配置文件 settings.py 的变量 MIDDLEWARE 中,以完成注册操作。

下面通过一个示例演示一下如何使用中间件。

(1) 创建名为 djangoProject 的 Django 项目。

(2) 打开命令提示符窗口,进入项目所在的根目录中,激活虚拟环境,并输入命令 python manage.py startapp book,创建名为 book 的应用,如图 5-195 所示。

图 5-195 创建应用

(3) 打开应用 book 中的文件 views.py,代码如下:

```python
# 资源包\Code\chapter5\5.11\1\djangoProject\book\views.py
from django.http import HttpResponse
def index(request):
    print('视图中执行的代码')
    return HttpResponse('项目运行成功!')
```

(4) 在应用 book 中创建自定义中间件文件 middlewares.py,代码如下:

```python
# 资源包\Code\chapter5\5.11\1\djangoProject\book\middlewares.py
# 函数
# def django_middleware(get_response):
#     print('自定义中间件初始化的相关代码')
#     def middleware(request):
#         print('请求到达视图之前所执行的代码')
#         response = get_response(request)
#         print('响应到达客户端浏览器之前所执行的代码')
#         return response
#     return middleware
# 类
class DjangoMiddleware(object):
    def __init__(self, get_response):
        print('自定义中间件初始化的相关代码')
        self.get_response = get_response
    def __call__(self, request):
        print('请求到达视图之前所执行的代码')
        response = self.get_response(request)
        print('响应到达客户端浏览器之前所执行的代码')
        return response
```

(5) 打开应用 djangoProject 中的文件 urls.py,代码如下:

```
# 资源包\Code\chapter5\5.11\1\djangoProject\djangoProject\urls.py
from django.urls import path
from book import views
urlpatterns = [
    path('', views.index),
]
```

（6）打开应用 djangoProject 中的配置文件 settings.py，对自定义中间件进行注册，代码如下：

```
# 资源包\Code\chapter5\5.11\1\djangoProject\djangoProject\settings.py
MIDDLEWARE = [
    'django.middleware.security.SecurityMiddleware',
    'django.contrib.sessions.middleware.SessionMiddleware',
    'django.middleware.common.CommonMiddleware',
    'django.middleware.csrf.CsrfViewMiddleware',
    'django.contrib.auth.middleware.AuthenticationMiddleware',
    'django.contrib.messages.middleware.MessageMiddleware',
    'django.middleware.clickjacking.XFrameOptionsMiddleware',
    'book.middlewares.DjangoMiddleware'
]
```

（7）运行上述项目，打开浏览器并在网址栏输入 http://127.0.0.1:8000/，其显示内容如图 5-196 所示。

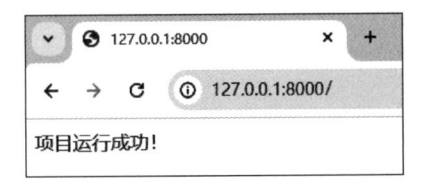

图 5-196　浏览器中的显示内容

此时，PyCharm 中的输出结果如图 5-197 所示。

图 5-197　PyCharm 中的输出结果

5.12　CSRF 防御

CSRF（Cross-Site Request Forgery，跨站请求伪造）是一种挟制用户在当前已登录的 Web 应用程序上执行非本意的操作的攻击方法。攻击者通过 HTTP 请求将数据发送到服务器，进而盗取 Cookie。在盗取到 Cookie 之后，攻击者不仅可以获取用户的相关信息，还可以修改该 Cookie 所关联的账户信息。

Django 提供了一套基于 Token 校验的完善的 CSRF 防护体系，仅需两步即可非常简单地解决 CSRF 攻击问题，一是开启 CSRF 中间件；二是在模板的表单当中添加模板变量{% csrf_token %}。

下面通过一个示例演示一下如何进行 CSRF 防御。

（1）创建名为 djangoProject 的 Django 项目。

（2）打开命令提示符窗口，进入项目所在的根目录中，激活虚拟环境，并输入命令 python manage.py startapp book，创建名为 book 的应用，如图 5-198 所示。

```
E:\Python全栈开发\djangoProject>workon django_env
(django_env) E:\Python全栈开发\djangoProject>python manage.py startapp book
```

图 5-198　创建应用

（3）打开应用 book 中的文件 views.py，代码如下：

```
# 资源包\Code\chapter5\5.12\1\djangoProject\book\views.py
from django.shortcuts import render
def index(request):
    return render(request, 'index.html')
```

（4）在项目根目录下的文件夹 templates 中创建模板文件 index.html，代码如下：

```
# 资源包\Code\chapter5\5.12\1\djangoProject\templates\index.html
<!DOCTYPE html>
<html lang="en">
    <head>
        <meta charset="UTF-8">
        <title>Title</title>
    </head>
    <body>
    <form action="">
        {% csrf_token %}
        <p>姓名：<input type="text"></p>
        <p><input type="submit" name="提交"></p>
    </form>
    </body>
</html>
```

（5）打开应用 djangoProject 中的文件 urls.py，代码如下：

```
# 资源包\Code\chapter5\5.12\1\djangoProject\djangoProject\urls.py
from django.urls import path
from book import views
urlpatterns = [
    path('', views.index),
]
```

（6）打开应用 djangoProject 中的配置文件 settings.py，确保 CSRF 中间件已开启。

（7）运行上述项目，打开浏览器并在网址栏输入 http://127.0.0.1:8000/，其显示内容如图 5-199 所示。

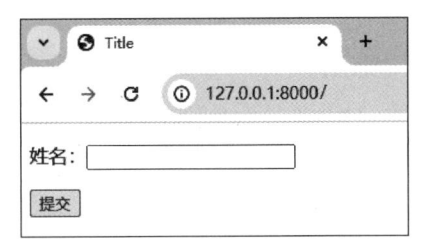

图 5-199　浏览器中的显示内容

（8）右击，单击"查看网页源代码"。此时，在 HTML 中创建了一个隐藏的字段，其中包含了 CSRF 令牌，其显示内容如图 5-200 所示。

图 5-200　浏览器中的显示内容

第 6 章 Django 项目实战：网上图书商城

本章将介绍如何实现网上图书商城，以便于更好地理解 Django 的相关使用方式。

6.1 程序概述

1．登录页面

登录页面主要用于客户账号和客户密码的输入，其界面如图 6-1 所示。

图 6-1　登录页面

2．注册页面

注册页面主要用于客户注册登录所需的账号，其界面如图 6-2 所示。

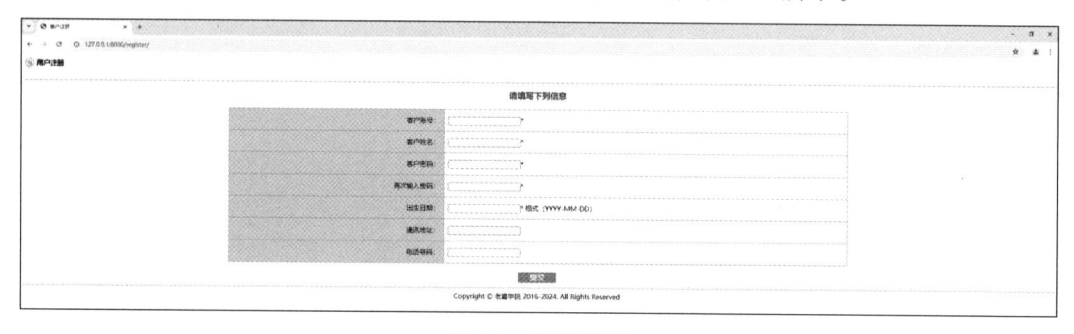

图 6-2　注册页面

3．主页面

主页面主要用于展示商品列表页面和购物车页面的导航，其界面如图 6-3 所示。

4．商品列表页面

商品列表页面主要用于显示全部商品，其界面如图 6-4 所示。

5．商品详情页面

商品详情页面主要用于显示商品的详细信息，其界面如图 6-5 所示。

图 6-3　主页面

图 6-4　商品列表页面

图 6-5　商品详情页面

6．购物车页面

购物车页面主要用于显示所购商品的订单信息，其界面如图 6-6 所示。

图 6-6　购物车页面

6.2　数据库设计

该项目的数据库中包含 4 个数据表，即客户表、商品表、订单表和详细订单表。

1．客户表

该表主要用于存储用户的相关信息，包括客户账号、客户姓名、客户密码、通信地址、电话号码和出生日期，其表结构如表 6-1 所示。

表 6-1　客户表的表结构

字 段 名	数 据 类 型	最 大 长 度	主　　键	外　　键	备　　注
id	CharField	20	YES	NO	客户账号
name	CharField	50	NO	NO	客户姓名

续表

字 段 名	数据类型	最大长度	主 键	外 键	备 注
password	CharField	20	NO	NO	客户密码
address	CharField	200	NO	NO	通信地址
phone	CharField	20	NO	NO	电话号码
birthday	DateField	—	NO	NO	出生日期

2. 商品表

该表主要用于存储商品的相关信息，包括图书名称、作者、出版社、ISBN、版次、包装、开本、出版时间、用纸、价格、图书详细描述和图书图片，其表结构如表 6-2 所示。

表 6-2　商品表的表结构

字 段 名	数据类型	最大长度	主 键	外 键	备 注
id	AutoField	—	YES	NO	图书 id
name	CharField	100	NO	NO	图书名称
author	CharField	100	NO	NO	作者
press	CharField	200	NO	NO	出版社
isbn	CharField	30	NO	NO	ISBN
edition	CharField	30	NO	NO	版次
packaging	CharField	30	NO	NO	包装
format	CharField	30	NO	NO	开本
publication_time	CharField	30	NO	NO	出版时间
paper	CharField	30	NO	NO	用纸
price	CharField	30	NO	NO	价格
description	CharField	200	NO	NO	图书详细描述
image	CharField	100	NO	NO	图书图片

3. 订单表

该表主要用于存储订单的相关信息，包括订单日期、订单付款状态和订单总价，其表结构如表 6-3 所示。

表 6-3　订单表的表结构

字 段 名	数据类型	最大长度	主 键	外 键	备 注
id	CharField	20	YES	NO	订单 id
order_date	DateTimeField	—	NO	NO	订单日期
status	IntegerField	—	NO	NO	订单付款状态
total	FloatField	—	NO	NO	订单总价

4. 详细订单表

该表主要用于存储详细订单的相关信息，包括图书 id、订单 id、图书数量和订单价格，其表结构如表 6-4 所示。

表 6-4　详细订单表的表结构

字 段 名	数据类型	长 度	主 键	外 键	备 注
id	AutoField	—	YES	NO	详细订单 id
goods	ForeignKey	—	NO	YES	图书 id

续表

字 段 名	数据类型	长 度	主 键	外 键	备 注
orders	ForeignKey	—	NO	YES	订单 id
quantity	IntegerField	—	NO	NO	图书数量
sub_total	FloatField	—	NO	NO	订单价格

6.3 编写程序

（1）创建名为 OnlineShoppingMall 的 Django 项目。

（2）打开命令提示符窗口，进入项目所在的根目录中，激活虚拟环境，并输入命令 python manage.py startapp book，创建名为 book 的应用，如图 6-7 所示。

```
E:\Python全栈开发\djangoProject>workon django_env
(django_env) E:\Python全栈开发\djangoProject>python manage.py startapp book
```

图 6-7　创建应用

（3）打开应用 book 中的文件 models.py，创建数据模型，代码如下：

```python
#资源包\OnlineShoppingMall\book\models.py
from django.db import models
#客户表
class Customer(models.Model):
    id = models.CharField(primary_key = True, max_length = 20)
    name = models.CharField(null = False, max_length = 50)
    password = models.CharField(null = False, max_length = 20)
    address = models.CharField(null = True, max_length = 200)
    phone = models.CharField(null = True, max_length = 20)
    birthday = models.DateField(null = True)
    class Meta:
        db_table = 'Customers'
        ordering = ['id']
#商品表
class Goods(models.Model):
    id = models.AutoField(primary_key = True)
    name = models.CharField(null = False, max_length = 100)
    author = models.CharField(null = False, max_length = 100)
    press = models.CharField(null = True, max_length = 200)
    isbn = models.CharField(null = True, max_length = 30)
    edition = models.CharField(null = True, max_length = 30)
    packaging = models.CharField(null = True, max_length = 30)
    format = models.CharField(null = True, max_length = 30)
    publication_time = models.CharField(null = True, max_length = 30)
    paper = models.CharField(null = True, max_length = 30)
    price = models.CharField(null = True, max_length = 30)
    description = models.CharField(null = True, max_length = 200)
    image = models.CharField(null = True, max_length = 100)
    class Meta:
        db_table = 'Goods'
        ordering = ['id']
#订单表
class Orders(models.Model):
```

```
    id = models.CharField(primary_key=True, max_length=20)
    order_date = models.DateTimeField()
    status = models.IntegerField(default=1)
    total = models.FloatField()
    class Meta:
        db_table = 'Orders'
        ordering = ['order_date']
# 订单详细表
class OrderLineItem(models.Model):
    id = models.AutoField(primary_key=True)
    goods = models.ForeignKey(Goods, on_delete=models.CASCADE)
    orders = models.ForeignKey(Orders, on_delete=models.CASCADE)
    quantity = models.IntegerField(default=0)
    sub_total = models.FloatField(default=0.0)
    class Meta:
        db_table = 'OrderLineItems'
        ordering = ['id']
```

（4）在项目的根目录下创建文件夹 static，用于存放静态文件。

（5）将文件夹 css、goods_images 和 images 存放至项目根目录下的文件夹 static 之中，如图 6-8 所示。

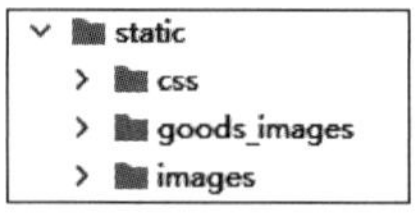

图 6-8 文件夹 static 的结构

（6）打开应用 OnlineShoppingMall 中的配置文件 settings.py，进行数据库配置、指定静态文件路径和应用注册等操作，代码如下：

```
# 资源包\OnlineShoppingMall\OnlineShoppingMall\settings.py
# 应用注册
INSTALLED_APPS = [
    'django.contrib.admin',
    'django.contrib.auth',
    'django.contrib.contenttypes',
    'django.contrib.sessions',
    'django.contrib.messages',
    'django.contrib.staticfiles',
    'book',
]
# 数据库配置
DATABASES = {
    'default': {
        # 数据库引擎
        'ENGINE': 'django.db.backends.mysql',
        # 数据库名称
        'NAME': 'django_db',
        # 连接 MySQL 的用户名
        'USER': 'root',
        # 连接 MySQL 的密码
        'PASSWORD': '12345678',
        # MySQL 的主机地址
        'HOST': '127.0.0.1',
        # MySQL 的端口号
        'PORT': '3306',
    }
}
# 指定静态文件路径
```

```
STATICFILES_DIRS = [
    os.path.join(BASE_DIR, "static")
]
```

(7) 在项目的根目录下的文件夹 templates 中创建父模板文件 base_header.html,用于显示带有统一标题的页面,代码如下:

```
# 资源包\OnlineShoppingMall\templates\base_header.html
<!doctype html>
{% load static %}
<html>
    <head>
        <meta charset="utf-8">
        <title>{% block title %}{% endblock %}</title>
        <link rel="stylesheet" type="text/css" href="{% static 'css/public.css' %}">
    </head>
    <body>
        <div class="header">网上图书商城</div>
        <hr width="100%" />
        {% block body %}{% endblock %}
        <div class="footer">
            <hr width="100%" />
            Copyright@ 老夏学院 2016-2024. All Rights Reserved
        </div>
    </body>
</html>
```

(8) 在项目根目录下的文件夹 templates 中创建父模板文件 base_title.html,用于显示带有自定义标题的页面,代码如下:

```
# 资源包\OnlineShoppingMall\templates\base_title.html
<!doctype html>
{% load static %}
<html>
    <head>
        <meta charset="utf-8">
        <title>{% block title %}{% endblock %}</title>
        <link rel="stylesheet" type="text/css" href="{% static 'css/public.css' %}">
    </head>
    <body>
        {% block body %}{% endblock %}
        <div class="footer">
            <hr width="100%" />
            Copyright@ 老夏学院 2016-2024. All Rights Reserved
        </div>
    </body>
</html>
```

(9) 在项目根目录下的文件夹 templates 中创建模板文件 login.html,用于显示客户登录页面,代码如下:

```
# 资源包\OnlineShoppingMall\templates\login.html
{% extends "base_header.html" %}
```

```
{% load static %}
{% block title %}客户登录{% endblock %}
{% block body %}
    <form action="/login/" method="post">
        {% csrf_token %}
        <table width="100%" align="center">
            <tr height="40">
                <td colspan="2" align="center"><strong>请您登录</strong></td>
            </tr>
            <tr height="40">
                <td width="50%" align="right"><img src="{% static 'images/3.jpg' %}" align="absmiddle"/>  客户账号：</td>
                <td>{{ form.userid }}</td>
            </tr>
            <tr height="40">
                <td width="50%" align="right"><img src="{% static 'images/2.jpg' %}" align="absmiddle"/>  客户密码：</td>
                <td>{{ form.password }}</td>
            </tr>
            <tr height="40">
                <td align="right"> </td>
                <td><input type="image" src="{% static 'images/login_button.jpg' %}"/><a href="/register/"><img src="{% static 'images/reg_button.jpg' %}" border="0"/></a></td>
            </tr>
        </table>
    </form>
{% endblock %}
```

(10) 在项目根目录下的文件夹 templates 中创建模板文件 customer_reg.html，用于显示客户注册页面，代码如下：

```
#资源包\OnlineShoppingMall\templates\customer_reg.html
{% extends "base_title.html" %}
{% load static %}
{% block title %}客户注册{% endblock %}
{% block body %}
    <style type="text/css">
        table {
            border-collapse: collapse;
        }
        .boder {
            border: 1px solid #5B96D0;
        }
        .col1 {
            background-color: #A6D2FF;
            text-align: right;
            padding-right: 10px;
            border: 1px solid #5B96D0;
            line-height: 50px;
        }
        .col2 {
            padding-left: 10px;
            border: 1px solid #5B96D0;
```

```
            line-height: 50px;
        }
        .textfield {
            height: 20px;
            width: 200px;
            border: 1px solid #999999;
            text-align: left;
            font-size: medium;
            line-height: 50px;
        }
</style>
<form action="/register/" method="post">
        {% csrf_token %}
        <div><img src="{% static 'images/reg.jpg' %}" align="absmiddle"/></div>
        <br>
        <hr width="100%"/>
        <ul>
            {% for field, errors in form.errors.items %}
                {% for message in errors %}
                    <li class="error">{{ message }}</li>
                {% endfor %}
            {% endfor %}
        </ul>
        <div class="text3" align="center">请填写下列信息</div>
        <br>
        <table width="60%" border="0" align="center" class="boder">
            <tr>
                <td width="35%" height="27" class="col1">客户账号：</td>
                <td class="col2">{{ form.userid }} *</td>
            </tr>
            <tr>
                <td height="27" class="col1">客户姓名：</td>
                <td class="col2">{{ form.name }} *</td>
            </tr>
            <tr>
                <td height="27" class="col1">客户密码：</td>
                <td class="col2">{{ form.password1 }} *</td>
            </tr>
            <tr>
                <td height="27" class="col1">再次输入密码：</td>
                <td class="col2">{{ form.password2 }} *</td>
            </tr>
            <tr>
                <td height="27" class="col1">出生日期：</td>
                <td class="col2">{{ form.birthday }} *
                    格式(YYYY-MM-DD)
                </td>
            </tr>
            <tr>
                <td height="27" class="col1">通信地址：</td>
                <td class="col2">{{ form.address }}</td>
            </tr>
            <tr>
                <td height="27" class="col1">电话号码：</td>
                <td class="col2">{{ form.phone }}</td>
            </tr>
```

```
        </table>
        <br>
        <div align = "center">
            <input type = "image" src = "{% static 'images/submit_button.jpg' %}"/>
        </div>
    </form>
{% endblock %}
```

(11) 在项目根目录下的文件夹 templates 中创建模板文件 customer_reg_success. html,用于显示客户注册成功后的页面,代码如下：

```
#资源包\OnlineShoppingMall\templates\customer_reg_success.html
{% extends "base_header.html" %}
{% block title %}注册成功{% endblock %}
{% block body %}
    <style type = "text/css">
        a:link {
            font-size: 18px;
            color: #DB8400;
            text-decoration: none;
            font-weight: bolder;
        }
        a:visited {
            font-size: 18px;
            color: #DB8400;
            text-decoration: none;
            font-weight: bolder;
        }
        a:hover {
            font-size: 18px;
            color: #DB8400;
            text-decoration: underline;
            font-weight: bolder;
        }
    </style>
    <div align = "center">
        <p class = "text7">恭喜您注册成功!</p>
        <p>
            <a href = "{% url 'login_view' %}">返回登录页面</a>
        </p>
    </div>
{% endblock %}
```

(12) 在项目根目录下的文件夹 templates 中创建模板文件 main. html,用于显示登录成功后的主页面,代码如下：

```
#资源包\OnlineShoppingMall\templates\main.html
{% extends "base_header.html" %}
{% load static %}
{% block title %}网上图书商城{% endblock %}
{% block body %}
    <style type = "text/css">
        a:link {
            font-size: 18px;
```

```css
        color: #DB8400;
        text-decoration: none;
        font-weight: bolder;
    }
    a:visited {
        font-size: 18px;
        color: #DB8400;
        text-decoration: none;
        font-weight: bolder;
    }
    a:hover {
        font-size: 18px;
        color: #DB8400;
        text-decoration: underline;
        font-weight: bolder;
    }
</style>
</head>
<div>
    <p class="text1"><img src="{% static 'images/4.jpg' %}" align="absmiddle"/><a href="/list/">商品列表</a></p>
    <p class="text2">您可以从产品列表中浏览感兴趣的产品进行购买</p>
</div>
<hr width="100%"/>
<div>
    <p class="text1"><img src="{% static 'images/mycar1.jpg' %}" align="absmiddle"/><a href="/show_cart/">购物车</a></p>
    <p class="text2">您可以把感兴趣的商品暂时放在购物车中</p>
</div>
{% endblock %}
```

（13）在项目根目录下的文件夹 templates 中创建父模板文件 goods_header.html，用于显示购物车、商品列表和注销等快捷链接，代码如下：

```html
#资源包\OnlineShoppingMall\templates\goods_header.html
{% load static %}
<td width="734" align="right">
    <img src="{% static 'images/mycar1.jpg' %}" align="absmiddle"/>
    <a href="/show_cart/"> 购物车</a> | <a href="/list/">商品列表</a>
    | <a href="/logout/">注销</a>
</td>
```

（14）在项目根目录下的文件夹 templates 中创建模板文件 goods_list.html，用于显示商品列表页面，代码如下：

```html
#资源包\OnlineShoppingMall\templates\goods_list.html
{% extends "base_title.html" %}
{% load static %}
{% block title %}商品列表{% endblock %}
{% block body %}
    <style type="text/css">
        table {
            border-collapse: collapse;
```

```
            }
            /* 商品列表第 1 列 */
            .col1 {
                padding - top: 5px;
                border - top: 1px dashed #666666;
                text - indent: 40px
            }
            /* 商品列表第 2 列 */
            .col2 {
                padding - top: 5px;
                border - top: 1px dashed #666666;
                text - align: right;
            }
            /* 商品列表第 3 列 */
            .col3 {
                padding - top: 5px;
                border - top: 1px dashed #666666;
                text - align: center;
            }
        </style>
        <table width = "100%" border = "0" align = "center">
            <tr>
                <td width = "616"><img src = "{% static 'images/list.jpg' %}" align = "absmiddle"/></td>
                {% include 'goods_header.html' %}
            </tr>
        </table>
        <hr width = "100%"/>
        <div class = "text3" align = "center">请从商品列表中选择您喜爱的商品</div>
        <br>
        <table width = "100%" border = "0" align = "center">
            <tr bgcolor = "#b4c8ed">
                <th>商品名称</th>
                <th width = "5%">商品价格</th>
                <th width = "15%">添加到购物车</th>
            </tr>
            {% for goods in goods_list %}
                <tr bgcolor = {% cycle '#ffffff' '#edf8ff' %}>
                    <td class = "col1"><a href = "/detail/?id = {{ goods.id }}">{{ goods.description }}</a></td>
                    <td class = "col2">¥{{ goods.price }}</td>
                    <td class = "col3"><a href = "/add/?id = {{ goods.id }}&name = {{ goods.name }}&price = {{ goods.price }}&page = list">添加到购物车</a></td>
                </tr>
            {% endfor %}
        </table>
{% endblock %}
```

(15) 在项目根目录下的文件夹 templates 中创建模板文件 goods_detail.html,用于显示商品详细信息页面,代码如下:

```
#资源包\OnlineShoppingMall\templates\goods_detail.html
{% extends "base_title.html" %}
{% load static %}
```

```html
{% block title %}客户注册{% endblock %}
{% block body %}
    <style type="text/css">
        .title {
            font-size: 20px;
            color: #FF6600;
            font-style: italic;
        }
    </style>
    <table width="100%" border="0" align="center">
        <tr>
            <td width="616"><img src="{% static 'images/info.jpg' %}" align="absmiddle"/></td>
            {% include 'goods_header.html' %}
        </tr>
    </table>
    <hr width="100%"/>
    <div class="text3" align="center">{{ goods.description }}</div>
    <table width="100%" border="0" align="center">
        <tr>
            <td width="40%" align="right">
                <div><img src="{% static 'goods_images/' %}{{ goods.image }}" width="360px" height="360px"/></div>
                <br></td>
            <td>
                <div align="center" class="text4">一 口 价:<span class="title">¥{{ goods.price }}元</span></div>
                <br>
                <table width="80%" height="200" border="0">
                    <tbody>
                        <tr>
                            <td width="25%" class="text5">作者:</td>
                            <td width="25%" class="text6">{{ goods.author }}</td>
                            <td width="25%" class="text5">出版社:</td>
                            <td width="25%" class="text6">{{ goods.press }}</td>
                        </tr>
                        <tr>
                            <td class="text5">ISBN:</td>
                            <td class="text6">{{ goods.isbn }}</td>
                            <td class="text5">版次:</td>
                            <td class="text6">{{ goods.edition }}</td>
                        </tr>
                        <tr>
                            <td class="text5">开本:</td>
                            <td class="text6">{{ goods.format }}</td>
                            <td class="text5">出版时间:</td>
                            <td class="text6">{{ goods.publication_time }}</td>
                        </tr>
                        <tr>
                            <td class="text5">用纸:</td>
                            <td class="text6">{{ goods.paper }}</td>
                            <td class="text5">包装:</td>
                            <td class="text6">{{ goods.packaging }}</td>
                        </tr>
                    </tbody>
                </table>
```

```
                    <br>
                    <br>
                    <div>
                        <a href="/add/?id={{ goods.id }}&name={{ goods.name }}&price={{ goods.price }}&page=detail">
                            <img src="{% static 'images/button.jpg' %}">
                        </a>
                    </div>
                </td>
            </tr>
    </table>
{% endblock %}
```

（16）在项目根目录下的文件夹 templates 中创建模板文件 cart.html，用于显示购物车页面，代码如下：

```
#资源包\OnlineShoppingMall\templates\cart.html
{% extends "base_title.html" %}
{% load static %}
{% block title %}购物车{% endblock %}
{% block body %}
    <style type="text/css">
        table {
            border-collapse: collapse;
        }
        .threeboder {
            border: 1px solid #5B96D0;
        }
        .trow {
            border-right: 1px solid #5B96D0;
            border-bottom: 1px solid #5A96D6;
        }
        .theader {
            background-color: #A5D3FF;
            font-size: 14px;
            border-right: 1px solid #5B96D0;
            border-bottom: 1px solid #5A96D6;
        }
    </style>
    <script>
        function calc(rowid, quantityInput) {
            quantity = quantityInput.value
            if (isNaN(quantity)) {
                alert("不是有效的数值!");
                quantityInput.value = 0;
                quantity = quantityInput.value
                quantityInput.focus();
            }
            //单价 id
            var price_id = 'price_' + rowid;
            //单价
            var price = parseFloat(document.getElementById(price_id).innerText);
            //小计 id
            var subtotal_id = 'subtotal_' + rowid;
```

```
                //小计(更新之前)
                subtotal1 = parseFloat(document.getElementById(subtotal_id).innerText);
                //四舍五入并保留两位小数
                subtotal1 = subtotal1.toFixed(2);
                document.getElementById(subtotal_id).innerText = quantity * price;
                //小计(更新之后)
                subtotal2 = parseFloat(document.getElementById(subtotal_id).innerText);
                //合计
                total = parseFloat(document.getElementById('total').innerText);
                //计算合计
                total = total - subtotal1 + subtotal2;
                //四舍五入并保留两位小数
                total = total.toFixed(2);
                //更新合计
                document.getElementById('total').innerText = total;
            }
        </script>
        <form action="/submit_orders/" method="post">
            {% csrf_token %}
            <table width="100%" border="0" align="center">
                <tr>
                    <td width="616"><img src="{% static 'images/mycar.jpg' %}"/></td>
                    {% include 'goods_header.html' %}
                </tr>
            </table>
            <hr width="100%"/>
            <div class="text3" align="center">您选好的商品</div>
            <br>
            <table width="100%" border="0" align="center" class="threeboder">
                <tr bgcolor="#A5D3FF">
                    <td height="50" align="center" class="theader">商品名称</td>
                    <td width="8%" align="center" class="theader">数量</td>
                    <td width="15%" align="center" class="theader">单价</td>
                    <td width="15%" align="center" class="theader">小计</td>
                </tr>
                {% for item in list %}
                <tr>
                    <td height="50" align="left" class="trow">  {{ item.1 }}</td>
                    <td align="center" class="trow">
                        <input name="quantity_{{ item.0 }}" type="text" value="{{ item.3 }}" onblur="calc({{ item.0 }}, this)"/>
                    </td>
                    <td align="center" class="trow">&yen;<span id="price_{{ item.0 }}">{{ item.2 }}</span></td>
                    <td align="center" class="trow">&yen;<span id="subtotal_{{ item.0 }}">{{ item.4 }}</span>
                    </td>
                </tr>
                {% endfor %}
                <tr>
                    <td height="50" colspan="5" align="right">合计: &yen;<span id="total">{{ total }}</span>  </td>
                </tr>
            </table>
            <br>
```

```html
        <div align="center">
            <a href="#"><input type="image" src="{% static 'images/submit_order.jpg' %}" border="0"/></a>  
        </div>
    </form>
{% endblock %}
```

(17) 在项目的根目录下的文件夹 templates 中创建模板文件 order_finish.html，用于显示订单完成页面，代码如下：

```html
#资源包\OnlineShoppingMall\templates\order_finish.html
{% extends "base_header.html" %}
{% block title %}订单完成{% endblock %}
{% block body %}
    <style type="text/css">
        a:link {
            font-size: 18px;
            color: #DB8400;
            text-decoration: none;
            font-weight: bolder;
        }
        a:visited {
            font-size: 18px;
            color: #DB8400;
            text-decoration: none;
            font-weight: bolder;
        }
        a:hover {
            font-size: 18px;
            color: #DB8400;
            text-decoration: underline;
            font-weight: bolder;
        }
    </style>
    <div align="center">
        <p class="text7">谢谢您的购物！</p>
        <p class="text7">您的订单号是：{{ ordersid }}</p>
        <p class="text7">您可以继续购物！</p>
        <p class="text7">
            <a href="/main/">返回主页面</a>
        </p>
    </div>
{% endblock %}
```

(18) 在应用 book 中创建文件 forms.py，用于对登录和注册信息进行验证，代码如下：

```python
#资源包\OnlineShoppingMall\book\forms.py
from django import forms
class LoginForm(forms.Form):
    userid = forms.CharField(label='客户账号：', required=True)
    password = forms.CharField(label='客户密码：', widget=forms.PasswordInput)
class RegistrationForm(forms.Form):
    userid = forms.CharField(label='客户账号：', required=True)
    name = forms.CharField(label='客户姓名：', required=True)
```

```python
        password1 = forms.CharField(label = '客户密码: ', widget = forms.PasswordInput)
        password2 = forms.CharField(label = '再次输入密码: ', widget = forms.PasswordInput)
        birthday = forms.DateField(label = '出生日期: ', error_messages = {'invalid': '输入的出生日期无效'})
        address = forms.CharField(label = '通信地址: ', required = False)
        phone = forms.CharField(label = '电话号码: ', required = False)
        def clean_password2(self):
            password1 = self.cleaned_data.get('password1')
            password2 = self.cleaned_data.get('password2')
            if password1 and password2 and password1 != password2:
                raise forms.ValidationError('两次输入的密码不一致')
            return password2
```

(19) 打开应用 book 中的 views.py 文件，用于编写登录、注册、主页面、商品列表、商品详细、添加购物车、显示购物车、提交订单、注销和初始化添加数据等视图函数，代码如下：

```python
# 资源包\OnlineShoppingMall\book\views.py
from django.shortcuts import render
from django.http import HttpResponse, HttpResponseRedirect
from django.views.generic import ListView
import random
import datetime
from book.forms import LoginForm, RegistrationForm
from book.models import Customer, Goods, OrderLineItem, Orders
# 登录
def login(request):
    if request.method == 'POST':
        form = LoginForm(request.POST)
        if form.is_valid():
            userid = form.cleaned_data['userid']
            password = form.cleaned_data['password']
            c = Customer.objects.get(id = userid)
            if c is not None and c.password == password:
                # 将客户 id 放到 Session 中
                request.session['customer_id'] = c.id
                return HttpResponseRedirect('/main/')
    else:
        form = LoginForm()
    return render(request, 'login.html', {'form': form})
# 注册
def register(request):
    if request.method == 'POST':
        form = RegistrationForm(request.POST)
        if form.is_valid():
            new_customer = Customer()
            new_customer.id = form.cleaned_data['userid']
            new_customer.name = form.cleaned_data['name']
            new_customer.password = form.cleaned_data['password1']
            new_customer.birthday = form.cleaned_data['birthday']
            new_customer.address = form.cleaned_data['address']
            new_customer.phone = form.cleaned_data['phone']
            new_customer.save()
            return render(request, 'customer_reg_success.html')
    else:
```

```python
        form = RegistrationForm()
    return render(request, 'customer_reg.html', {'form': form})
#主页面
def main(request):
    #判断客户是否已经登录
    if not request.session.has_key('customer_id'):
        return HttpResponseRedirect('/login/')
    return render(request, 'main.html')
#商品列表
class GoodsListView(ListView):
    model = Goods
    ordering = ['id']
    template_name = 'goods_list.html'
#商品详细
def show_goods_detail(request):
    goodsid = request.GET['id']
    goods = Goods.objects.get(id=goodsid)
    return render(request, 'goods_detail.html', {'goods': goods})
#添加购物车
def add_cart(request):
    #判断客户是否已经登录
    if not request.session.has_key('customer_id'):
        return HttpResponseRedirect('/login/')
    goodsid = int(request.GET['id'])
    goodsname = request.GET['name']
    goodsprice = float(request.GET['price'])
    #判断Session中是否已经存在购物车数据
    if not request.session.has_key('cart'):
        #如果没有,则创建一个空购物车,购物车采用的是列表结构
        request.session['cart'] = []
    cart = request.session['cart']
    #声明一个标志位,0表示购物车中没有当前商品,1表示购物车中有当前商品
    flag = 0
    for item in cart:
        #item采用的也是列表结构,例如[商品id,商品名称,价格,数量]
        if item[0] == goodsid:
            item[3] += 1
            flag = 1
            break
    if flag == 0:
        item = [goodsid, goodsname, goodsprice, 1]
        cart.append(item)
    request.session['cart'] = cart
    print(cart)
    page = request.GET['page']
    if page == 'detail':
        return HttpResponseRedirect('/detail/?id=' + str(goodsid))
    else:
        return HttpResponseRedirect('/list/')
#显示购物车
def show_cart(request):
    #判断客户是否已经登录
    if not request.session.has_key('customer_id'):
        return HttpResponseRedirect('/login/')
    if not request.session.has_key('cart'):
        print('购物车是空的')
```

```python
        return render(request, 'cart.html', {'list': [], 'total': 0.0})
    cart = request.session['cart']
    list = []
    total = 0.0
    for item in cart:
        #item 结构 [商品 id, 商品名称, 价格, 数量]
        #计算小计
        subtotal = item[2] * item[3]
        total += subtotal
        #将小计追加到 new_item (商品 id, 商品名称, 价格, 数量, 小计)
        new_item = (item[0], item[1], item[2], item[3], subtotal)
        list.append(new_item)
    return render(request, 'cart.html', {'list': list, 'total': total})
#提交订单
def submit_orders(request):
    if request.method == 'POST':
        orders = Orders()
        #生成订单 id,规则是当前时间戳+1位随机数
        n = random.randint(0, 9)
        d = datetime.datetime.today()
        ordersid = str(int(d.timestamp() * 1e6)) + str(n)
        orders.id = ordersid
        orders.order_date = d
        orders.status = 1
        orders.total = 0.0
        orders.save()
        cart = request.session['cart']
        total = 0.0
        for item in cart:
            #item 结构:[商品 id, 商品名称, 价格, 数量]
            goodsid = item[0]
            goods = Goods.objects.get(id=goodsid)
            quantity = request.POST['quantity_' + str(goodsid)]
            try:
                quantity = int(quantity)
            except:
                quantity = 0
            #计算小计
            subtotal = item[2] * quantity
            total += subtotal
            order_line_item = OrderLineItem()
            order_line_item.quantity = quantity
            order_line_item.goods = goods
            order_line_item.orders = orders
            order_line_item.sub_total = subtotal
            order_line_item.save()
        orders.total = total
        orders.save()
        #提交订单后购物车应清除
        del request.session['cart']
        return render(request, 'order_finish.html', {'ordersid': ordersid})
#注销
def logout(request):
    if request.session.has_key('customer_id'):
        del request.session['customer_id']
        if request.session.has_key('cart'):
```

```
            del request.session['cart']
        return HttpResponseRedirect('/login/')
# 初始化添加数据
def add_goods(request):
    Goods.objects.bulk_create([Goods(1, 'Python全栈开发——基础入门', '夏正东', '清华大学出
版社', '9787302600909', '1',
                                    '平装', '16 开', '2022 - 07 - 01', '胶版纸', '79',
                                    'Python全栈开发——基础入门(清华开发者书库.Python)
作者:夏正东 Python 畅销书籍',
                                    'dc670c75c629548c.jpg'),
                              Goods(2, 'Python全栈开发——高阶编程', '夏正东', '清华大学
出版社', '9787302608943', '1',
                                    '平装', '16 开', '2022 - 08 - 01', '胶版纸', '89',
                                    'Python全栈开发——高阶编程(清华开发者书库.Python)
作者:夏正东 Python 畅销书籍',
                                    '14ce4d2acd51eed8.jpg'),
                              Goods(3, 'Python全栈开发——数据分析', '夏正东', '清华大学
出版社', '9787302625001', '1',
                                    '平装', '16 开', '2023 - 03 - 01', '胶版纸', '79',
                                    'Python全栈开发——数据分析(清华开发者书库.Python)
作者:夏正东 Python 畅销书籍',
                                    '6130ebb1f9e2c229.jpg')])
    return HttpResponse('已成功添加多册图书!')
```

(20)打开应用 OnlineShoppingMall 中的文件 urls.py,代码如下:

```
# 资源包\OnlineShoppingMall\OnlineShoppingMall\urls.py
from django.urls import path, re_path
from book import views
urlpatterns = [
    re_path('^$|^login/$', views.login, name = 'login_view'),
    path('logout/', views.logout, name = 'logout_view'),
    path('register/', views.register, name = 'register_view'),
    path('main/', views.main, name = 'main_view'),
    path('list/', views.GoodsListView.as_view(), name = 'list_view'),
    path('detail/', views.show_goods_detail, name = 'detail_view'),
    path('add/', views.add_cart),
    path('show_cart/', views.show_cart, name = 'cart_view'),
    path('submit_orders/', views.submit_orders),
    path('add_goods/', views.add_goods),
]
```

(21)打开应用 OnlineShoppingMall 中的文件__init__.py,代码如下:

```
# 资源包\OnlineShoppingMall\OnlineShoppingMall\__init__.py
import pymysql
pymysql.install_as_MySQLdb()
```

(22)打开命令提示符窗口,进入项目所在的根目录中,激活虚拟环境,并输入命令 python manage.py makemigrations,生成迁移脚本文件,如图 6-9 所示。

(23)再次输入命令 python manage.py migrate,将迁移脚本文件映射到数据库中,如图 6-10 所示。

(24)运行上述项目,打开浏览器并在网址栏输入 http://127.0.0.1:8000/add_

图 6-9　生成迁移脚本文件

图 6-10　将迁移脚本文件映射到数据库中

goods/，添加初始化数据，其显示内容如图 6-11 所示。

图 6-11　浏览器中的显示内容

（25）在浏览器中的网址栏输入 http://127.0.0.1:8000/，即可正常访问网站。

图书推荐

书 名	作 者
仓颉语言实战(微课视频版)	张磊
仓颉语言核心编程——入门、进阶与实战	徐礼文
仓颉语言程序设计	董昱
仓颉程序设计语言	刘安战
仓颉语言元编程	张磊
仓颉语言极速入门——UI全场景实战	张云波
HarmonyOS移动应用开发(ArkTS版)	刘安战、余雨萍、陈争艳 等
公有云安全实践(AWS版·微课视频版)	陈涛、陈庭暄
虚拟化KVM极速入门	陈涛
虚拟化KVM进阶实践	陈涛
移动GIS开发与应用——基于ArcGIS Maps SDK for Kotlin	董昱
Vue+Spring Boot前后端分离开发实战(第2版·微课视频版)	贾志杰
前端工程化——体系架构与基础建设(微课视频版)	李恒谦
TypeScript框架开发实践(微课视频版)	曾振中
精讲MySQL复杂查询	张方兴
Kubernetes API Server源码分析与扩展开发(微课视频版)	张海龙
编译器之旅——打造自己的编程语言(微课视频版)	于东亮
全栈接口自动化测试实践	胡胜强、单镜石、李睿
Spring Boot+Vue.js+uni-app全栈开发	夏云虎、姚晓峰
Selenium 3自动化测试——从Python基础到框架封装实战(微课视频版)	栗任龙
Unity编辑器开发与拓展	张寿昆
跟我一起学uni-app——从零基础到项目上线(微课视频版)	陈斯佳
Python Streamlit从入门到实战——快速构建机器学习和数据科学Web应用(微课视频版)	王鑫
Java项目实战——深入理解大型互联网企业通用技术(基础篇)	廖志伟
Java项目实战——深入理解大型互联网企业通用技术(进阶篇)	廖志伟
深度探索Vue.js——原理剖析与实战应用	张云鹏
前端三剑客——HTML5+CSS3+JavaScript从入门到实战	贾志杰
剑指大前端全栈工程师	贾志杰、史广、赵东彦
JavaScript修炼之路	张云鹏、戚爱斌
Flink原理深入与编程实战——Scala+Java(微课视频版)	辛立伟
Spark原理深入与编程实战(微课视频版)	辛立伟、张帆、张会娟
PySpark原理深入与编程实战(微课视频版)	辛立伟、辛雨桐
HarmonyOS原子化服务卡片原理与实战	李洋
鸿蒙应用程序开发	董昱
HarmonyOS App开发从0到1	张诏添、李凯杰
Android Runtime源码解析	史宁宁
恶意代码逆向分析基础详解	刘晓阳
网络攻防中的匿名链路设计与实现	杨昌家
深度探索Go语言——对象模型与runtime的原理、特性及应用	封幼林
深入理解Go语言	刘丹冰
Spring Boot 3.0开发实战	李西明、陈立为

续表

书 名	作 者
全解深度学习——九大核心算法	于浩文
HuggingFace自然语言处理详解——基于BERT中文模型的任务实战	李福林
动手学推荐系统——基于PyTorch的算法实现(微课视频版)	於方仁
深度学习——从零基础快速入门到项目实践	文青山
LangChain与新时代生产力——AI应用开发之路	陆梦阳、朱剑、孙罗庚、韩中俊
图像识别——深度学习模型理论与实战	于浩文
编程改变生活——用PySide6/PyQt6创建GUI程序(基础篇·微课视频版)	邢世通
编程改变生活——用PySide6/PyQt6创建GUI程序(进阶篇·微课视频版)	邢世通
编程改变生活——用Python提升你的能力(基础篇·微课视频版)	邢世通
编程改变生活——用Python提升你的能力(进阶篇·微课视频版)	邢世通
Python量化交易实战——使用vn.py构建交易系统	欧阳鹏程
Python从入门到全栈开发	钱超
Python全栈开发——基础入门	夏正东
Python全栈开发——高阶编程	夏正东
Python全栈开发——数据分析	夏正东
Python编程与科学计算(微课视频版)	李志远、黄化人、姚明菊 等
Python数据分析实战——从Excel轻松入门Pandas	曾贤志
Python概率统计	李爽
Python数据分析从0到1	邓立文、俞心宇、牛瑶
Python游戏编程项目开发实战	李志远
Java多线程并发体系实战(微课视频版)	刘宁萌
从数据科学看懂数字化转型——数据如何改变世界	刘通
Dart语言实战——基于Flutter框架的程序开发(第2版)	亢少军
Dart语言实战——基于Angular框架的Web开发	刘仕文
FFmpeg入门详解——音视频原理及应用	梅会东
FFmpeg入门详解——SDK二次开发与直播美颜原理及应用	梅会东
FFmpeg入门详解——流媒体直播原理及应用	梅会东
FFmpeg入门详解——命令行与音视频特效原理及应用	梅会东
FFmpeg入门详解——音视频流媒体播放器原理及应用	梅会东
FFmpeg入门详解——视频监控与ONVIF+GB28181原理及应用	梅会东
Python玩转数学问题——轻松学习NumPy、SciPy和Matplotlib	张骞
Pandas通关实战	黄福星
深入浅出Power Query M语言	黄福星
深入浅出DAX——Excel Power Pivot和Power BI高效数据分析	黄福星
从Excel到Python数据分析:Pandas、xlwings、openpyxl、Matplotlib的交互与应用	黄福星
云原生开发实践	高尚衡
云计算管理配置与实战	杨昌家
HarmonyOS从入门到精通40例	戈帅
OpenHarmony轻量系统从入门到精通50例	戈帅
AR Foundation增强现实开发实战(ARKit版)	汪祥春
AR Foundation增强现实开发实战(ARCore版)	汪祥春